Driver Reactions to Automated Vehicles

A Practical Guide for Design and Evaluation

Transportation Human Factors: Aerospace, Aviation, Maritime, Rail, and Road Series

Series Editor
Professor Neville A. Stanton
University of Southampton, UK

Driver Reactions to Automated Vehicles
A Practical Guide for Design and Evaluation

By

Alexander Eriksson and Neville A. Stanton

CRC Press
Taylor & Francis Group
Boca Raton London New York

CRC Press is an imprint of the
Taylor & Francis Group, an **informa** business

MATLAB® is a trademark of The MathWorks, Inc. and is used with permission. The MathWorks does not warrant the accuracy of the text or exercises in this book. This book's use or discussion of MATLAB® software or related products does not constitute endorsement or sponsorship by The MathWorks of a particular pedagogical approach or particular use of the MATLAB® software.

CRC Press
Taylor & Francis Group
6000 Broken Sound Parkway NW, Suite 300
Boca Raton, FL 33487-2742

© 2018 by Taylor & Francis Group, LLC
CRC Press is an imprint of Taylor & Francis Group, an Informa business

No claim to original U.S. Government works

Printed on acid-free paper

International Standard Book Number-13: 978-0-8153-8282-9 (Hardback)

Visit the Taylor & Francis Web site at
http://www.taylorandfrancis.com

and the CRC Press Web site at
http://www.crcpress.com

For my family – Alexander

For Maggie, Josh and Jemima – Neville

Contents

Preface

The motivation for this book came to us because we wanted to communicate the research we have been conducting as part of a project on the Human Factors of Highly Automated Vehicles. This research has been conducted in driving simulators and on test tracks as well as the open road. Road vehicle automation is a fast-moving field, with manufacturers developing their own versions, such as the Audi A8 Traffic Jam Pilot, BMW Smartcruise, Cadillac Super Cruise, Mercedes Distronic Plus and Tesla Autopilot. On the face of it, these technologies seem to offer many benefits to drivers, freeing them up to do other, non-driving, tasks. However, in reality, these technologies require drivers to remain attentive to the automated driving system and other vehicles around them. Ultimately, the driver may be called to take over control from the vehicle, as the automation cannot cope with the road conditions and/or complexity of the environment. It is also possible that the sensors may fail or conflict with each other.

To address these problems, we looked at automation in other domains to understand why the human-automation system may not always result in anticipated benefits. Our research showed that the communication between automation and humans may fall short of ideal requirements. This lesson was taken into consideration in our own investigations. First of all, we developed software so we could recreate vehicle automation in our own driving simulator. This was no trivial undertaking, and now this software is being used by other researchers, including those working for vehicle manufacturers. Then, we investigated the handover from driver to vehicle and back again. What we found here is that the time taken to perform the handover (in non-emergency situations) was much longer than previously reported in the literature. In addition, the handover from driver to vehicle tends not to be reported. We also found that self-paced handover from vehicle to driver results in greater driving stability than reported in emergency handovers.

We have also performed one of the very few studies of vehicle–driver handover of control on open roads (the M40 motorway in the United Kingdom). This study was undertaken in a Tesla Model S P90 equipped with the Autopilot software feature. Interestingly, we found that the handover times, from driver to vehicle and vice versa, correlated very highly with the handover times from our simulator studies (although drivers were, on average, about one second quicker on the open road). This gives us increased confidence for using driving simulators in studies of vehicle automation. Our findings from these and other studies led us to design new interfaces between the automation and the driver, which we tested in a driving simulator. The interfaces that allowed the driver to plan ahead, supporting tactical level by aiding decision-making, showed the best performance.

It is our hope that the content of this book will help to guide engineers, system designers and researchers through the complexities of human–automation interaction (in particular for the design of handovers). We have demonstrated methods for studies of performance in simulators and on open roads. We have also raised issues surrounding the complexity of bringing the human driver back into control of the vehicle. Vehicle automation is already upon us, but the issues are far from resolved. There is much work to be done before vehicle automation becomes commonplace, and the field needs to move quickly.

MATLAB® is a registered trademark of The MathWorks, Inc. For product information, please contact:

The MathWorks, Inc.
3 Apple Hill Drive
Natick, MA 01760-2098 USA
Tel: 508 647 7000
Fax: 508-647-7001
E-mail: info@mathworks.com
Web: www.mathworks.com

Authors

Alexander Eriksson, PhD, is a researcher at the Swedish National Road and Transport Research Institute (VTI) in Gothenburg, Sweden. He is also the competence area leader for driving simulator application at SAFER Vehicle and Traffic Safety Centre at Chalmers. He has an MSc and a BSc in cognitive science from Linköping University, Sweden, and received his PhD within the Marie Curie ITN HF-Auto from the Faculty of Engineering and the Environment at the University of Southampton in the United Kingdom, where he is currently a visiting research fellow. During his PhD studies, he was seconded to Jaguar Land Rover and the Technical University Munich (TUM). He was also tasked with running the Southampton University Driving Simulator lab facility during his PhD research. His research interests lie within traffic safety, human performance, vehicle automation and how interaction between humans and technical systems may be facilitated.

Professor Neville Stanton, PhD, DSc, is a chartered psychologist, chartered ergonomist and chartered engineer. He holds the Chair in Human Factors Engineering in the Faculty of Engineering and the Environment at the University of Southampton in the United Kingdom. He has degrees in occupational psychology, applied psychology and human factors engineering and has worked at the Universities of Aston, Brunel, Cornell and MIT. His research interests include modelling, predicting, analysing and evaluating human performance in systems as well as designing the interfaces and interaction between humans and technology. Professor Stanton has worked on design of automobiles, aircraft, ships and control rooms over the past 30 years, on a variety of automation projects. He has published 40 books and over 300 journal papers on ergonomics and human factors. In 1998, he was presented with the Institution of Electrical Engineers Divisional Premium Award for research into System Safety. The Institute of Ergonomics and Human Factors in the United Kingdom awarded him the Otto Edholm Medal in 2001, the President's Medal in 2008 and the Sir Frederic Bartlett Medal in 2012 for his contributions to basic and applied ergonomics research. The Royal Aeronautical Society awarded him and his colleagues the Hodgson Prize in 2006 for research on design-induced, flight-deck error, published in *The Aeronautical Journal*. The University of Southampton awarded him a Doctor of Science in 2014 for his sustained contribution to the development and validation of human factors methods.

Acknowledgements

The authors would like to acknowledge a number of people who have been instrumental to the research presented in this book. First, we would like to thank our funding agency, The European Union's Marie Skłodowska-Curie actions for making this research possible as a part of the HFAuto project (PITN-GA-2013-605817). We would also like to extend our gratitude to Jaguar Land Rover, in particular Jim O'donoghue, and to the Technical University Munich (TUM), who greatly contributed to parts of the research presented in this book through the hosting and facilitation of Dr. Eriksson's and Prof. Stanton's research. There are also some key researchers who contributed to the research presented within parts of this book, amongst others Dr. Victoria A. Steane, Dr. Sebastiaan Petermeijer and Dr. Joost C. F. De Winter contributed significantly by carrying out parts of the research presented in this book. The authors have thoroughly enjoyed working and collaborating with the these researchers and institutes.

Needless to say, it is difficult to imagine such successful research findings without the great friends and collaborators who share a passion for research.

Lastly, we would like to extend our gratitude to our families for the support provided throughout our research careers.

Definitions and Abbreviations

Absolute validity is achieved when there is a perfect correspondence between results obtained in the simulator and in the real world.

ACC adaptive cruise control

ADAS advanced driver assistance systems

ADS automated driving system

AOA angle of attack

AOI area of interest

ARM altitude pre-select

BASt bundesanstalt für straβwesen

BEA bureau d'enquêtes et d'analyses pour la sécurité de l'aviation civile

CG common ground

COCOM contextual control model

CVR cockpit voice recorder

CWP control wheel position

DA driving Automation

DDT dynamic driving task

ECAM electronic centralised aircraft monitor

FDR flight data recorder

Fidelity can be divided into three types: equipment, environmental and psychological. Equipment fidelity describes the degree to which the simulator corresponds to the physical layout of a vehicle; environmental describes the feel and appearance of the simulator (vehicle dynamics, controller feel and behaviour); and environmental fidelity describes to what degree the simulator replicates the sensory stimulation of driving a real vehicle, for example, through visual feedback, moving platforms, linear actuators or rumble tables.

FL flight level

HMI human–machine interfaces

LWD left wing down

MSL mean sea level

NHTSA national highway traffic safety administration

NTSB national transportation safety board

ODD operational design domain

ODI office of defects investigations

OEDR object and event detection and response

PF pilot flying

PNF pilot not flying

Relative validity is achieved when there is a lack of a perfect correspondence between the results obtained in a simulator and the real world but the effect has got the same direction or relative size in the simulator as in real-world conditions.

RWD	right wing down
SA	situation awareness
SAE	society of automotive engineers
SDSWA	standard deviation of steering wheel angle
THW	time-headway
TORlt	take-over request lead time
TOrt	take-over request reaction time
TTC	time to collision
Validity	describes whether the results obtained in a simulator correspond to results obtained in a real environment. Can be divided into relative and absolute validity.

1 Introduction

1.1 BACKGROUND

Humans have used wheeled modes of transport to shuttle people and goods ever since the invention of the wheel in the late Neolithic era (4500–3300 BCE). The modes of transport were limited to carriages drawn by large, strong animals until the invention of the lightweight, high-pressure steam engine in 1797 by Richard Trevithick (Nokes, 1899). This enabled large volumes of goods to be shipped on 'railed' carriages. Less than 100 years later (1886), Carl Benz showcased his 'motor car' (Figure 1.1), thereby introducing the first ever automobile (Benz, 1886), replacing the horse and carriage as a daily mode of transport.

Following the invention of the first motor vehicle, development of vehicle technologies moved quickly, from the introduction of comfort systems through the invention of the automatic transmission (Figure 1.2) by Alfred Horner in 1921 (Alfred, 1923), to safety features such as the energy-absorbing steering column developed by Daimler in 1967 (Skeels, 1966). In 1972, the first 'computer' made it into a car in the form of the Buick Max Trac traction control system for the Riviera Silver Arrow III (Nattrass, 2015).

However, the notion of autonomous cars was born more than 30 years earlier at the 1939 World Fair, where GM showcased a concept of the Automated Roadway of the Future (Urmson et al., 2008). Not much later, in 1956, Central Power and Light Company posted an advert (Figure 1.3) stating, 'One day your car may speed along an electric super-highway, its speed and steering automatically controlled by electronic devices embedded in the road'. It is evident, therefore, that the notion of cars that drive themselves predates the introduction of computing technology in vehicles.

As computing technology evolved in the coming years, fuelled by Moore's law (Moore, 1965), the vision of the self-driving car came closer to reality. An early proof-of-concept of intelligent transport systems was requested by the US government in 1991 to reduce fatal road accidents which at the time, in the United States, amounted to 40,000 fatalities and a cost of $150 billion on a yearly basis. The proof-of-concept did not receive funding until 1995 and was anticipated to be presented by 1997 (Thorpe et al., 1997). In 1997, researchers from a Carnegie Mellon University (CMU) research group demonstrated their Free Agent Scenario, which could be described as a first-generation Society of Automotive Engineers (SAE) Level 3 vehicle demonstrator. The demonstrator consisted of two 'fully automated' Pontiac Bonneville sedans, a partially automated Oldsmobile Silhouette minivan and two fully automated New Flyer city buses (Thorpe et al., 1997).

All vehicles were running the CMU-developed Rapid Adapting Lateral Position Handler (RALPH) vision system, as well as radars from Delco and Eaton Vorad and Light Detection and Ranging devices (LIDAR) from Rigel (Thorpe et al., 1997). Each vehicle drove approximately 64–68 drives on a proving ground in San

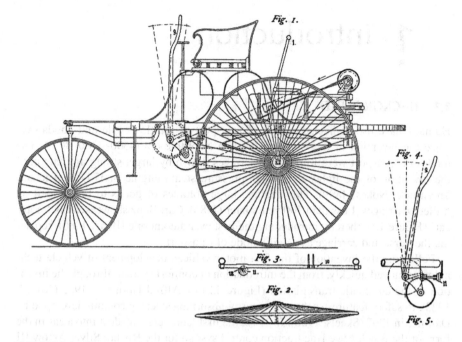

FIGURE 1.1   The patent for the first ever automobile.

Diego during the course of five weeks demonstrating features, such as platooning, obstacle avoidance, intelligent lane changes and overtaking manoeuvres (Thorpe et al., 1997). However, this was not the earliest large demonstration of 'self-driving' technologies. Two years earlier, in 1995, two researchers from CMU drove hands-free for 2797 miles from Washington, DC, to San Diego in their prototype vehicle using the RALPH vision system (Jochem et al., 1995; Port, 1995). This is a particularly impressive feat given that Delphi made a similar (3400 mile) trip from San Francisco to New York completely hands- and feet-free in 2015. The difference between these endeavours was that the Delphi vehicle was equipped with 'six long-range radars, four short-range radars, three vision-based cameras, six LIDAR's, a localization system, intelligent software algorithms and a full suite of Advanced Drive Assistance Systems' (Delphi, 2017), whereas RALPH merely had one camera (440×525 tv-lines) facing the roadway, and about 1/10th of the computing power and just over 1/20th of the RAM in the first-generation Apple Watch. The Delphi coast-to-coast drive is not the only recent effort in demonstrating long journeys driven by automated driving systems. In 2010, four fully automated vehicles drove 13,000 kilometres from Italy to Shanghai as part of the VisLab Intercontinental Autonomous Challenge (Broggi et al., 2012), using vehicles based on the BRAin-drIVE (BRAiVE) platform, equipped with '10 cameras, 4 laser scanners, 16 laser beams, 1 radar, a GPS and an Inertial Measurement Unit' (2012, p. 148) as well as an additional laser scanner. In 2012, Google reported that their fully autonomous vehicle fleet had reached the 500,000-kilometre mark (Urmson, 2012).

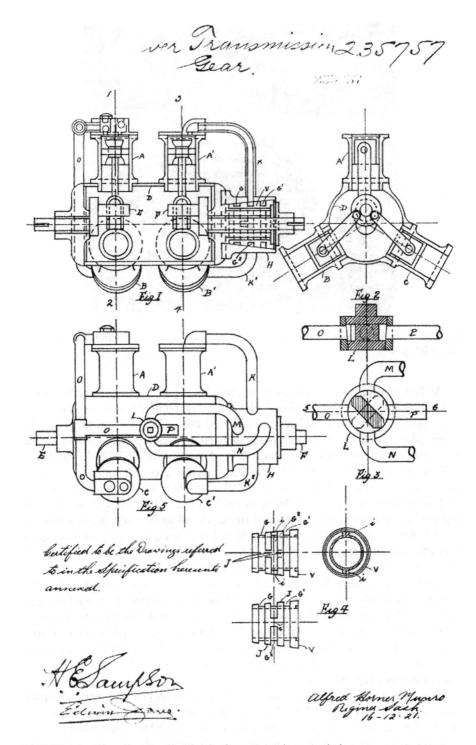

FIGURE 1.2 The patent drawing for the first automatic transmission.

FIGURE 1.3   An America's Power Companies advertisement from 1956, featuring futuristic self-driving cars.

At first glance, it may seem as if progress has been stagnant since the 1990s. Contemporary automated driving technologies encountered some of the same challenges faced by the CMU designers of RALPH, namely that the driver is still expected to resume control when the automated driving system is no longer able to handle a situation. This is currently defined by the SAE in Table 1.1 as 'Level 2/3' automation (SAE J3016, 2016). Indeed, much like contemporary systems by manufacturers such as Tesla, Google, BMW, Volvo and Audi (to name a few) rely on having the driver act as a fallback, so did RALPH: 'When the system can't find regular lane markings, such as painted stripes or reflecting studs, it can key in on other features: the edge of the road, dark lines left by tires, even the trail of oil spots down the center of most lanes. When all such signs vanish, Ralph shakes the wheel and beeps, alerting the human driver to take over' (Port, 1995).

Regardless of the technological advances in the automotive industry allowing drivers to be hands- and feet-free through the use of automation, it has until recently been a legal requirement that drivers must remain in overall control of their vehicle at all times, as stipulated in Article 8 of the Vienna Convention on Road Traffic (United Nations, 1968). This requirement has since been amended in the convention, now stating that drivers can be hands- and feet-free, as long as the system can be shut off or overridden by the driver.

Consequently, drivers of contemporary vehicles equipped with basic autopilot features (e.g. Mercedes S-Class, Tesla) are now able to legally relinquish control of the driving task to manufacturer proprietary algorithms, transitioning from a driver monitoring, to a driver not driving (Banks et al., in press; Banks and

TABLE 1.1

A Description of the (SAE J3016, 2016) Levels of Automated Driving

Level of Automation		Narrative Description as Defined in SAE J3016
0	No driving automation	The performance by the driver of the entire DDT[a], even when enhanced by active safety systems
1	Driver assistance	The sustained and ODD[a]-specific execution by a driving automation system of either the lateral or the longitudinal vehicle motion control subtask of the DDT (but not both simultaneously) with the expectation that the driver performs the remainder of the DDT
2	Partial driving automation	The sustained and ODD-specific execution by a driving automation system of both the lateral and longitudinal vehicle motion control subtasks of the DDT with the expectation that the driver completes the OEDR[a] subtask and supervises the driving automation system
3	Conditional driving automation	The sustained and ODD-specific performance by an ADS[a] of the entire DDT with the expectation that the DDT fallback-ready user is receptive to ADS-issued requests to intervene and to DDT performance-relevant system failures in other vehicle systems, and will respond appropriately
4	High driving automation	The sustained and ODD-specific performance by an ADS of the entire DDT and DDT fallback without any expectation that a user will respond to a request to intervene
5	Full driving automation	The sustained and unconditional (i.e. not ODD-specific) performance by an ADS of the entire DDT and DDT fallback without any expectation that a user will respond to a request to intervene

[a] ODD, operational design domain; DDT, dynamic driving task; ADS, automated driving System; OEDR, object and event detection and response.

Stanton, in press), from human-in-the-loop to human-on-the-loop (Nothwang et al., 2016), or from a tactical (attended) to a tactical (unattended) level of control (as further described in Chapter 6 and Hollnagel and Woods, 2005). This becomes quite problematic, as some of the marketing for SAE Level 2 systems implies functionality to that of SAE Level 3 (Phys. Org, 2016), indicating that these vehicles will be able to notify the driver when the vehicle decides it can no longer handle a situation. This 'dissonance' was recently demonstrated by the fatal crash of a Tesla Model S (Office of Defects Investigation, 2017), where, on the 7th of May 2015, a Tesla Model S collided with a tractor's trailer crossing an intersection on a highway west of Williston, Florida, causing fatal harm to the driver of the Tesla. Data from Tesla indicates that, at the time of the incident, the vehicle in question was being operated in autopilot mode. Based on this, it is evident that the limitations of such Level 2/3 systems must be conveyed to the drivers as to avoid disuse and misuse of the technology (Eriksson and Stanton, 2017a; Parasuraman and Riley, 1997), and that ways of facilitating the transition between manual and automated vehicle control must be explored at this level of automation to allow this technology to save rather than take lives.

1.2 RESEARCH MOTIVATION

In 2015, the World Health Organization reported that 1.2 million people lost their lives in vehicular accidents (World Health Organization, 2015). This is about 2150 times more than in aviation during the same time-period. Claims made by Elon Musk, CEO of Tesla Motors, one of the manufacturers that were early to market with semi-automated vehicles, state that 'The probability of having an accident is 50% lower if you have Autopilot on. Even with our first version' (Musk, 2016).

If this claim is correct, and driver assistance systems such as these become a standard feature of next-generation vehicles, it would not only save millions of lives worldwide, but also help the European Commission reach the goal of reducing road traffic deaths by 50% by 2020 (European Commission, 2010). It has been estimated that over 90% of crashes are caused by the driver (Singh, 2015; Thorpe et al., 1997). It must be noted that consumer adoption of automated vehicles will not happen overnight. Consequently, a transition period is to be expected where automated and manually driven vehicles share the same infrastructure (Ioannou, 1998).

This has raised some safety concerns, as automated vehicles will have to be able to predict the behaviour of a somewhat unpredictable human driver, whilst drivers of manual vehicles must anticipate and adapt to the driving style of automated vehicles seeking to optimise performance (van Loon and Martens, 2015). Whilst an immediate uptake and substantial reduction in accident rates may not be feasible at low uptake rates of automated vehicles, there are still significant benefits to be reaped from uptake rates as low as 2%–5%. Such benefits include improved traffic flow and a reduction in phantom traffic jams (Fountoulakis et al., 2017; Ioannou, 1998). However, as uptake rates increase, it is anticipated that vehicle accidents will decline; consequently, this technology would not only save human lives, but it could also contribute to significant reductions in medical and emergency service costs through the reduction in accident rates, allowing for funds allocated to such services to be allocated elsewhere.

Not only do automated vehicles offer a reduction in road accidents, but they also offer significant reductions in congestion due to shorter inter-vehicular spacing as well as reduced pollution and fuel consumption (Fagnant and Kockelman, 2015; Kyriakidis et al., 2017; Willemsen et al., 2015). If these gains outweigh the potential cost of developing the technology and acquiring vehicles equipped with this technology, then automation may be found beneficial in economic, societal and environmental terms (Stanton and Marsden, 1996; Young et al., 2011). However, recent estimates state that automated vehicles need to be driven between 1.6 million and 11 billion miles (depending on the safety thresholds), taking up to 500 years for a fleet of 100 vehicles, to statistically assess the safety of automated driving technology (Kalra and Paddock, 2016).

Despite the recent evolution of automated vehicles, concerns have been expressed that just because something can be automated, does not mean it should be (Fitts, 1951; Hancock, 2014). Whilst contemporary automated (SAE Level 2) driving allows hands-free and feet-free driving for short periods of time, it still relies on the driver to act as a fallback when the automated driving system can no longer handle a situation. This is despite well-established knowledge that humans are poor monitors

(Parasuraman, 1987), and are prone to misuse, disuse or even abuse of automation technology (Parasuraman and Riley, 1997). This means that Human Factors researchers must work to ensure safe usage of this technology, and to ensure efficient communication of the limitations of automated driving systems (Eriksson and Stanton, 2017a), until such systems reach a level of performance (SAE Level 4 and above) where the driver is no longer a critical part of the driving task (SAE J3016, 2016).

1.3 RESEARCH AIMS

Whilst automated vehicles are portrayed as a panacea in road safety and environmental gains, there is a concern amongst the Human Factors community that automated vehicle systems on the SAE intermediary level (Level 1–3) are likely to be the cause for some issues. In 1983, Elisabeth Bainbridge (1983) published a seminal paper detailing the 'ironies of automation'. Such ironies included poor monitoring. It is well established that it is impossible for a human to maintain visual attention to a source of information where very little happens for more than half an hour (Mackworth, 1948), and in automated vehicles, drivers seem to lose interest in vehicle performance after a very short time (Banks et al., 2018). This is problematic as contemporary systems rely on the driver acting as a fallback when system boundaries are approached. Whilst contemporary research has focussed on transitions from automated to manual vehicle control based on 'system-limits', such as sensor failures or unexpected conditions, there is a paucity of research into driver-paced transitions of control. Driver-paced transitions of control are of utmost importance, as the self-regulation of a process often results in a higher net performance on said task (Eriksson et al., 2014; Hollnagel and Woods, 2005). Indeed, Bainbridge states that:

> Non-time-stressed operators, if they find they have the wrong type of display, might themselves request a different level of information. This would add to the work load of someone making decisions which are paced by a dynamic system.

(Bainbridge, 1983, pp. 778)

This research attempts to address concerns over ascertaining the control transition process in automated vehicles. The overarching goal of this book is to establish how drivers resume control from automated vehicles in non-urgent situations, and how such transitions can be facilitated by allowing the drivers to pace the transition, and whether the transition can be facilitated through the use of feedback in the human–machine interface. In order to address this, four sub-goals have been posited.

1. Proposing a framework for reasoning about human–automation interaction, based on linguistic theories into human–human communication.

   It is important to have a framework to reason about the interaction between driver and vehicle, as appropriate communication is deemed to be of utmost importance for safety in automated systems. Basing such a framework on the principles used when humans interact amongst each other could help facilitate the interaction between humans and automated agents in the intermediary stages of automated driving where the driver is expected to be available for manual control.

2. Designing Open Source software algorithms that will allow a generic driving simulator to drive automated, in a way that enables dynamic driver interaction with the system.

   Supplying a set of Open Source algorithms for automated driving working out-of-the-box on the STISIM driving simulation platform, as well as being easily implemented on other driving simulation platforms, allows the Human Factors community to assess dynamic driver behaviour without risking driver safety on the road as the algorithms have been validated against on-road conditions for research on control transition. Moreover, a consistent set of 'publicly-available' algorithms will enable consistency in the approach to study automated driving in the simulator, thus enabling generalisability across results from different simulators and researchers.

3. Validate the Southampton University Driving Simulator and the algorithms in Chapter 3 against driver behaviour during real-world driving conditions.

   Validating the Southampton University Driving Simulator's capabilities for automated driving research against on-road driving conditions is a requirement to ensure that the results presented in this book are comparable to that of real driving conditions and that the data generated by the simulator holds any value in assessing driver behaviour. Indeed, according to Rolfe et al., 'The value of a simulator depends on its ability to elicit from the operator the same sort of response that he would make in the real situation' (Rolfe et at., 1970, p. 761). Moreover, establishing validity of the Southampton University Driving Simulator for automated driving research will be of great value for forthcoming projects utilising the simulator for automated driving research.

4. Provide design guidance for vehicle manufacturers, and guidelines for policymakers on the transition process between automated and manual driving based on experimental evidence.

   The provision of design guidance for vehicle manufacturers is important as it strives to ensure that the driver–vehicle interaction is facilitated to its furthest extent. Moreover, providing guidelines and information based on scientific findings to policymakers is of utmost importance as policy will dictate the overall functional requirements of automated vehicles in terms of facilitating driver interaction in the intermediary levels of automation (e.g. ascertaining the control transition process in urgent and non-urgent transitions of control).

1.4 OUTLINE OF THE BOOK

1.4.1 CHAPTER 1: INTRODUCTION

This initial chapter provides a short history of automated vehicles and introduces the area of human factors in highly automated vehicles. It outlines the taxonomy of levels of automation used throughout this book as defined by SAE as well as the objectives of the research conducted as part of this book, along with a chapter-by-chapter summary and the book's overall contribution to knowledge.

1.4.2 CHAPTER 2: A LINGUISTICS APPROACH FOR ASSESSING HUMAN–AUTOMATION INTERACTION

This chapter examines previous research regarding human–automation interaction and the challenges caused by the introduction of automation in complex systems. One of the challenges identified is the increased effort to interpret the plethora of additional information supplied by introducing automation into the driving task. In an effort to decrease the effort to interpret system states, this chapter draws on theories from human–human communication to highlight the requirement to make such systems transparent and bridge the gap between determining system states and how the system behaviour corresponds to expectations.

To achieve this, the chapter posits a novel application of linguistics and, more specifically, the Gricean theories of human–human communication (Grice, 1975) to the domain of human–automation interaction, treating the automated driving system as an intelligent agent or a co-driver whose task is to ensure that the driver does not have an incorrect model of the vehicle state, is aware of changes to the driving dynamics and has the necessary knowledge to perform the driving task. The proposed application of linguistic theory to human–automation interaction is then exemplified through the application of the framework to two incidents in aviation where automation played a significant role in the events that unfolded.

1.4.3 CHAPTER 3: A TOOLBOX FOR AUTOMATED DRIVING RESEARCH IN THE SIMULATOR

Driving simulators are a common and important tool used to conduct research in the automotive domain. This chapter details the value of using simulators as an alternative to on-road testing, as it provides a safe, fast and cheap way of researching human behaviour in vehicles, and with automated driving systems. Moreover, this chapter describes the design and implementation of a set of generic automated driving algorithms and the performance of said algorithms in terms of car-following and lane-keeping performance. This chapter also describes the algorithms as implemented on the STISIM v3 driving simulation platform as part of the experimental set-up used in Chapters 6 and 8.

1.4.4   CHAPTER 4: TAKE-OVER TIME IN HIGHLY AUTOMATED VEHICLES

Through a review of the contemporary literature into control transitions describing response-times ranging from 2 to 3.5 seconds in critical situations with varying lead times, a knowledge vacuum pertaining to the control transition process in higher levels of automation (SAE Level 4) was identified. Thus, this chapter examines how long drivers take to resume control in non-critical situations from a highly automated vehicle in two conditions, with and without a secondary task, when prompted to resume control in a driving simulator. This chapter found that automating the driving task has a detrimental effect on driver reaction time. Adequate reaction times are crucial for the feasibility of the lower levels of automation (SAE Level 2–3) as drivers are expected to be able to intervene at a moment's notice, meaning that the process of resuming control is of utmost importance to study. These results are then contrasted with the results described in the 25 papers included in the review of the literature.

Moreover, a further knowledge gap was identified as there are no reports in the contemporary literature regarding how long it takes drivers to transition from manual, to automated driving modes, something that is crucial when designing vehicle platooning. Thus, the chapter addresses this by determining a range of times that drivers took to make the transition from manual to automated driving.

1.4.5   CHAPTER 5: CONTRASTING SIMULATED WITH
ON-ROAD TRANSITIONS OF CONTROL

During the research conducted for Chapter 4, it was found that no research into control transition had taken place on public roads, and that contemporary on-road research in automation either focussed on sub-systems such as adaptive cruise control or other automated features on test tracks. Consequently, there is very little research being done on the open road that can serve as a validation of the findings from research carried out in simulators worldwide. In an effort to establish validity for research on control transitions in the simulator, the Southampton University Driving Simulator and the algorithms in Chapter 3, this chapter examines whether the findings from Chapter 4 could be transferred to open-road driving conditions. A between-group comparison of the data collected for Chapter 4 in the simulator using the algorithms from Chapter 3 and a group of drivers driving a Tesla Model S with the autopilot feature was carried out. The driving task in the on-road element of the study was replicated to the furthest extent to match the passive monitoring condition used in Chapter 4. The results showed strong correlations between the distributions of control transition times from manual to automated, as well as automated to manual control. Consequently, it can be concluded that the driving simulator is a valid research tool for studying human factors of automated vehicles and lends validity to research into control transitions carried out in the simulator already disseminated in the scientific literature. Moreover, the findings in this chapter lend validity to the algorithms presented in Chapter 3 as well as the Southampton University Driving Simulator.

1.4.6 CHAPTER 6: AFTER-EFFECTS OF DRIVER-PACED TRANSITIONS OF CONTROL

As relative validity between the driving simulator and on-road driving could be established for the control transition task, further analysis of the after-effects of driving performance data was carried out. This was motivated by contemporary literature into control transitions (reported on in Chapter 4, Table 4.1) reporting that detrimental effects on driving performance could be seen in all but one study observing close to normal driving performance when a lead time of 8 s was used (Damböck et al., 2012).

This led to the hypothesis that drivers should exhibit little detrimental effects on their driving performance after resuming control from an automated vehicle, and that their driving performance should not change when resuming control from the vehicle after being engaged in a secondary task, as the drivers were able to moderate the transition process through the time they took to resume control. The findings show that there are no large effects of self-paced control transitions, contrary to what has been reported in the literature when a take-over request (TOR) was issued and control was transferred under time pressure, or without pre-emptive warnings of the need for driver intervention. This further strengthens the case made in Chapter 5 that transitions of control should be driver-paced when possible.

1.4.7 CHAPTER 7: AUGMENTED REALITY GUIDANCE FOR CONTROL TRANSITIONS IN AUTOMATED VEHICLES

This chapter explores how drivers can be supported through transitions of control by means of feedback through the in-vehicle human–machine interface when a situation requires the driver to make a tactical choice (i.e. slowing down or changing lanes to adapt to changing road conditions). Based on the findings in Chapter 2, a number of interfaces were assessed, categorised into three of the four levels of automation proposed by Parasuraman et al. (2000). In the experiment, the driver was given a request to resume control six times in four conditions, where four levels of information were presented as in a heads-up display as augmented reality. The effect of the interfaces was assessed from a performance perspective; that is, whether optimal actions are implemented and how different levels of information acquisition affect gaze-scanning behaviour, which is directly related to the safe manual handling of a road vehicle. That the user interfaces assessed in this chapter would help with the initial driver response to the take-over request was neither expected nor found. However, significant improvements appeared when assessing the effect of the user interface in the cognitive processing and action selection phase, lending support to the findings of Chapter 2 stating that information should be succinctly presented in a salient and relevant location in order to not present oversaturated information to the driver causing a decline in driving and task performance.

1.4.8 CHAPTER 8: CONCLUSIONS AND FUTURE WORK

The final chapter summarises the objectives set out in Chapter 1 in light of the experimental findings presented in this book and reflects on the contributions made to knowledge. An assessment of the approach taken in this book, by evaluating driver-paced tasks rather than focussing on system-paced transitions of control, as in the contemporary literature, showcases the importance of considering human variability and flexibility.

This insight may provide researchers with an alternative method complementing fully controlled research studies, where behaviour is highly influenced by the experimental design, with a semi-controlled approach that captures more naturalistic behaviours. This chapter also reflects on the theoretical framework, the validation of simulators as a tool and the practical implications of the research conducted as part of this book. Finally, areas of further enquiry are identified.

1.5 CONTRIBUTION TO KNOWLEDGE

The primary contribution of this book is achieved through the increased understanding of how humans interact with automated systems and the process in which manual control is resumed from an automated vehicle. The book has laid out a theoretical, linguistics-based framework for reasoning about human–automation interaction. This framework was then partly assessed in Chapter 7, where a head-up augmented reality human–machine interface was assessed in terms of decision support, lending evidence to the theoretical arguments laid out in Chapter 2, regarding using Paul Grice (1975) maxims as a means of assessing communication between entities in human–agent systems.

The book then assessed the process in which control is transferred from the vehicle to the driver. Whilst the topic is well researched – for a review see Eriksson and Stanton (2017b) and Chapter 4 – a lack of knowledge regarding transitions of control in non-critical conditions was identified. This was addressed in Chapters 4 through 6, more specifically in terms of transition time variability, validity and whether the after-effects observed in contemporary literature on the topic of control transitions could be found. These findings provide further support for Hollnagel and Woods (2005) Contextual Control Model (COCOM) which identified time as a crucial component in determining the resulting control state and control performance. Higher levels of control in the self-paced research studies was found compared with those reported in the contemporary literature. Moreover, it is anticipated that the evidence presented in this book will help serve policymakers in the design and outlining of policy for contemporary and future highly automated vehicles, and it will serve as a source of knowledge and information for vehicle manufacturers to draw upon when designing the highly automated vehicles of the future.

This book provides a methodological contribution by disseminating the design and Open Source implementation of a set of algorithms that enable dynamic driver interaction with automated driving features in simulators. These algorithms will allow users of STISIM (or other simulators) to conduct research on a platform that can be accessed by any user of STISIM, thus allowing for the direct comparison of findings between research groups.

Furthermore, these algorithms have been experimentally validated against on-road driving conditions in a Tesla Model S during the task of transitioning between control modes, not only lending validity to the algorithms in Chapter 3 as a research tool, but also lending validity to previous research on the assessment of control transitions in the simulator. In addition to validating the algorithms in Chapter 3, the book also provides validation for the Southampton University Driving Simulator with regard to task validity. The results from Chapter 4 and Chapter 5 indicate high task validity and thus relative validity for the task of transferring control between manual driving and automated driving in the Southampton University Driving Simulator when correlated to on-road driving conditions in a Tesla Model S equipped with SAE Level 2 automation ($r = 0.97$ for transitions to manual control from automated driving, and $r = 0.96$ for transitions from manual to automated vehicle control). The validation of the Southampton University Driving Simulator for automated driving research is of utmost importance as it generates confidence in the results presented not only in this book, but from forthcoming projects utilising the simulator for automated driving research as well.

REFERENCES

Alfred, H. M. (1923). Power transmission gear: Google patents.

Bainbridge, L. (1983). Ironies of automation. *Automatica*, vol 19, no 6, pp 775–779.

Banks, V. A., Eriksson, A., O'Donoghue, J., Stanton, N. A. (2018). Is partially automated driving a bad idea? Observations from an on-road study. *Applied Ergonomics*, vol 68, pp 138–145.

Banks, V. A., and Stanton, N. A. (In press). Analysis of driver roles: Modelling the changing role of the driver in automated driving systems using EAST. *Theoretical Issues in Ergonomics Science*. Retrieved from https://doi.org/10.1080/1463922X.2017.1305465.

Benz, K. F. (1886). Fahrzeug mit Gasmotorenbetrieb. *German Patent Luft-und Gaskraftmaschinen*, no 37435.

Broggi, A., Cerri, P., Felisa, M., Laghi, M. C., Mazzei, L., and Porta, P. P. (2012). The VisLab Intercontinental Autonomous Challenge: An extensive test for a platoon of intelligent vehicles. *International Journal of Vehicle Autonomous Systems*, vol 10, no 3, pp 147–164.

Damböck, D., Bengler, K., Farid, M.,, Tönert, L. (2012). Übernahmezeiten beim hochautomatisierten Fahren. *Tagung Fahrerassistenz München*, vol 15, pp 16.

Delphi. (2017). Delphi drive. *Delphi* Retrieved 30 March, 2017.

Eriksson, A., Lindström, A., Seward, A., Seward, A., Kircher, K. (2014). Can user-paced, menu-free spoken language interfaces improve dual task handling while driving? In M. Kurosu (Ed.), *Human-Computer Interaction. Advanced Interaction Modalities and Techniques* (vol 8511). Cham, Switzerland: Springer, pp 394–405.

Eriksson, A.,Stanton, N. A. (2017a). The chatty co-driver: A linguistics approach applying lessons learnt from aviation incidents. *Safety Science*, vol 99, pp 94–101.

Eriksson, A., Stanton, N. A. (2017b). Takeover time in highly automated vehicles: Noncritical transitions to and from manual control. *Human Factors*, vol 59, no 4, pp 689–705.

European Commission. (2010). *Towards a European R oad S afety A rea: Policy O rientations on R oad S afety 2011–2020*. European Commission.

Fagnant, D. J., Kockelman, K. (2015). Preparing a nation for autonomous vehicles: Opportunities, barriers and policy recommendations. *Transportation Research Part a-Policy and Practice*, vol 77, pp 167–181.

Fitts, P. M. (Ed.) (1951). Human engineering for an effective air-navigation and traffic-control system. Oxford, England: National Research Council.

Fountoulakis, M., Bekiaris-Liberis, N., Roncoli, C., Papamichail, L., Papageorgiou, M. (2017). Highway traffic state estimation with mixed connected and conventional vehicles: Microscopic simulation-based testing. *Transportation Research Part C-Emerging Technologies*, vol 78, pp 13–33.

Grice, H. P. (1975). Logic and conversation. In P. Cole and J. L. Morgan (Eds.), *Speech Acts* (pp 41–58). New York, NY: Academic Press.

Hancock, P. A. (2014). Automation: How much is too much? *Ergonomics*, vol 57, no 3, pp 449–454.

Hollnagel, E., Woods, D. D. (2005). *Joint Cognitive Systems Foundations of Cognitive Systems Engineering*. Boca Raton, FL: CRC Press.

Ioannou, P. (1998). Evaluation of mixed automated/manual traffic. California Partners for Advanced Transit and Highways (PATH). Final Report MOU#290 ISSN 1055-142.

Jochem, T., Pomerleau, D., Kumar, B., and Armstrong, J. (1995). PANS: A portable navigation platform. Paper presented at the Intelligent Vehicles' 95 Symposium, Detroit, MI, pp 107–112.

Kalra, N., and Paddock, S. M. (2016). Driving to safety: How many miles of driving would it take to demonstrate autonomous vehicle reliability? *Transportation Research Part A: Policy and Practice*, vol 94, pp 182–193.

Kyriakidis, M., de Winter, J. C. F., Stanton, N., Bellet, T., van Arem, B., Brookhuis, K., Martens, M. H., Bengler, K., Andersson, J., Merat, N., Reed, N., Flament, M., Hagenzieker, M., and Happee, R. (2017). A human factors perspective on automated driving. *Theoretical Issues in Ergonomics Science*, pp 1–27.

Mackworth, N. H. (1948). The breakdown of vigilance during prolonged visual search. *Quarterly Journal of Experimental Psychology*, vol 1, no 1, pp 6–21.

Moore, G. E. (1965). Cramming more components onto integrated circuits. *Electron*, vol 38, pp 114–117.

Musk, E. (2016). Interview at 'Future Transport Solutions' conference. Retrieved from https://youtube/HaJAF4tQVbA?t=1341 Quote at 22 minutes 36 seconds: retrieved on 12 May 2016.

Nattrass, K. (2015). Buick's influential five – concepts that drove the brand's design and technology; inspired the industry. Retrieved from http://media.buick.com/media/us/en/buick/vehicles/concepts/avenir/2015.detail.html/content/Pages/news/us/en/2015/Jan/naias/buick/0111-buick-inf-5.html.

Nokes, G. A. (1899). *The Evolution of the Steam Locomotive (1803 to 1898)*. London: The Railway Publishing Company.

Nothwang, W. D., McCourt, M. J., Robinson, R. M., Burden, S. A., Curtis, J. W. (2016). The human should be part of the control loop? Paper presented at the Resilience Week (RWS), Chicago, IL.

Office of Defects Investigation. (2017). Automatic vehicle control systems. *National Highway Traffic Safety Administration*.

Parasuraman, R. (1987). Human-computer monitoring. *Human Factors*, vol 29, no 6, pp 695–706.

Parasuraman, R., Riley, V. (1997). Humans and automation: Use, misuse, disuse, abuse. *Human Factors*, vol 39, no 2, pp 230–253.

Parasuraman, R., Sheridan, T. B., Wickens, C. D. (2000). A model for types and levels of human interaction with automation. *IEEE Transactions on Systems, Man, and Cybernetics Part A, Systems and Humans*, vol 30, no 3, pp 286–297.

Phys.Org. (2016, October 17, 2016). Berlin tells Tesla: Stop ads with 'misleading' autopilot term. Retrieved from https://phys.org/news/2016-10-berlin-tesla-ads-autopilot-term.html.

Port, O. (1995). Look ma, no hands. Bloomberg. Retrieved from https://www.bloomberg.com/news/articles/1995-08-13/look-ma-no-hands.

Rolfe, J. M., Hammerton-Fraser, A. M., Poulter, R. F., Smith, E. M. (1970). Pilot response in flight and simulated flight. *Ergonomics*, vol 13, no 6, pp 761–768.

SAE J3016. (2016). Taxonomy and Definitions for Terms Related to Driving Automation Systems for On-Road Motor Vehicles, J3016_201609: SAE International.

Singh, S. (2015). Critical reasons for crashes investigated in the national motor vehicle crash causation survey. No. DOT HS 812 115. A Brief Statistical Summary, February.

Skeels, P. C. (1966). The general motors energy-absorbing steering column. Paper presented at the SAE Technical Paper Series. Retrieved from http://dx.doi.org/10.4271/660785.

Stanton, N. A., Marsden, P. (1996). From fly-by-wire to drive-by-wire: Safety implications of automation in vehicles. *Safety Science*, vol 24, no 1, pp 35–49.

Thorpe, C., Jochem, T., Pomerleau, D. (1997). The 1997 automated highway free agent demonstration. Paper presented at the IEEE Conference on Intelligent Transportation System, ITSC'97, Boston, MA.

United Nations. (1968). *Convention on Road Traffic*, Vienna, 8 November 1968. Amendment 1. Retrieved from http://www.unece.org/fileadmin/DAM/trans/conventn/crt1968e.pdf.

Urmson, C. (2012). The self-driving car logs more miles on new wheels. *Google official blog*.

Urmson, C., Duggins, D., Jochem, T., Pomerleau, D., Thorpe, C. (2008). From automated highways to urban challenges. Paper presented at the IEEE International Conference on Vehicular Electronics and Safety, ICVES 2008, Columbus, OH.

van Loon, R. J., Martens, M. H. (2015). Automated driving and its effect on the safety ecosystem: How do compatibility issues affect the transition period? *Procedia Manufacturing*, vol 3, pp 3280–3285.

Willemsen, D., Stuiver, A., Hogema, J. (2015). Automated driving functions giving control back to the driver: A simulator study on driver state dependent strategies. Paper presented at the 24th International Technical Conference on the Enhanced Safety of Vehicles (ESV), Gothenburg, Sweden.

World Health Organization. (2015). Global Health Observatory (GHO) data: Number of road traffic deaths. Retrieved from http://www.who.int/gho/road_safety/mortality/number_text/en/.

Young, M. S., Birrell, S. A., and Stanton, N. A. (2011). Safe driving in a green world: A review of driver performance benchmarks and technologies to support 'smart' driving. *Applied Ergonomics*, vol 42, no 4, pp 533–539.

2 A Linguistics Approach for Assessing Human–Automation Interaction

2.1 INTRODUCTION

Driving automation (DA) involves the automation of one or more higher-level cognitive driving tasks, such as maintaining longitudinal and/or lateral vehicle position in relation to traffic and road environments (Young et al., 2007). DA distinguishes itself from vehicle automation by entailing forms of automation that involve the psychological part of driving, namely the tactical, operational and strategic levels of driving (Michon, 1985). The higher levels of control involving complex decisions and planning would qualify as DA (Young et al., 2007). By using DA in highly automated vehicles, all but the strategic level of driving could be transferred from the driver to the DA system. Only the highest level of control, that is, goal setting on a strategic level, would remain with the driver for the main part of the journey. According to the Society of Automotive Engineers (SAE) International and the Bundesanstalt für Straßenwesen (BASt), DA functionality is likely to be limited to certain geographical areas, such as motorways (Gasser et al., 2009; SAE J3016, 2016), until full autonomy (SAE Level 5) can be achieved. Thus, there is a need for a human driver whose task is to resume control of the vehicle when the operational limits of DA are approached (Hollnagel and Woods, 2005; Stanton et al., 1997). This use of DA fundamentally alters the driving task (Hollnagel and Woods, 1983; Parasuraman et al., 2000; Woods, 1996) and will likely give rise to automation surprises (Sarter et al., 1997) and ironies (Bainbridge, 1983), such as an unevenly distributed workload (Hollnagel and Woods, 1983; Hollnagel and Woods, 2005; Kaber and Endsley, 1997; Kaber et al., 2001; Norman, 1990; Parasuraman, 2000; Sheridan, 1995; Woods, 1993; Young and Stanton, 1997, 2002, 2007b), loss of situation awareness (SA) and poor vigilance (Endsley, 1996; Endsley et al., 1997; Kaber and Endsley, 1997; Kaber and Endsley, 2004; Kaber et al., 2001; Sheridan, 1995; Woods, 1993) with the risk of ending up out-of-the-loop (Endsley, 1996; Endsley et al., 1997; Kaber and Endsley, 1997; Kaber and Endsley, 2004; Kaber et al., 2001; Norman, 1990), as well as the possibility of mode errors (Andre and Degani, 1997; Degani et al., 1995; Leveson, 2004; Norman, 1983; Rushby et al., 1999; Sarter and Woods, 1995; Sheridan, 1995).

These problems manifest when the driver is required to return to the driving control loop, either due to mechanical malfunctions, sensor failure or when the vehicle approaches a context where automation is no longer supported, such as adverse weather conditions, adverse behaviour of other road users or unforeseen events in the road environment. An example of a contextual restriction in contemporary DA is adaptive cruise control (ACC). Using ACC for prolonged periods of time may cause drivers to forget that the ACC system is still engaged when it is time to leave the motorway, which, in busy traffic where vehicle speed is limited by other road users, could result in an increase of vehicle velocity when taking an off-ramp as there are no vehicles in front of the car (Norman, 2009). It is, therefore, important to ensure that the driver receives the support and guidance necessary to safely get back into the vehicle control loop (Cranor, 2008).

Failure-induced transfer of control has been extensively studied (see Chapter 4; Desmond et al., 1998; Molloy and Parasuraman, 1996; Stanton et al., 1997; Strand et al., 2014; Young and Stanton, 2007a). It takes approximately 1 second for a driver to respond to a sudden braking event in traffic (Summala, 2000; Swaroop and Rajagopal, 2001; Wolterink et al., 2011). A technical failure leading to an unplanned and immediate transfer of vehicle control back to the driver will likely give rise to an incident as the 0.3 second time-headway (the time between the leading and host vehicle as a function of velocity and distance) is shorter than driver response-times (Willemsen et al., 2015). Given that drivers are unlikely to be able to intervene in situations where a response-time of less than 1 second is required, it is arguable that the likelihood of failure-induced transfer of control must be made negligible. The feasibility of DA rests on the systems' ability to cope with all but the most severe technical failures without loss of control on public roads.

Routine transfers of control under 'normal' circumstances have not been studied as extensively as failure-induced transfers of control (see Chapter 4 and Eriksson and Stanton, 2017). Therefore, many factors still need to be explored, such as what method and time is used to transfer control, how the human–machine interface will convey necessary information, and how the transfer of control will be managed by the driver (Beiker, 2012; Hoc et al., 2009; Merat et al., 2014). Christoffersen and Woods (2002) stated that in order to ensure coordination between human and machine, the system state must be transparent enough for the agents to understand problems and activities, as well as the plans of other agents, and how they cope with external events such as traffic and sensor disturbances (Beller et al., 2013; Inagaki, 2003; Kaber et al., 2001; Klein et al., 2004; Rankin et al., 2013; Weick et al., 2005).

This decreases the size of what Norman (2013) refers to as the gulf of evaluation, which is the effort required to interpret the state of the system and determine how well the behaviour corresponds to expectations. This puts a requirement on designers and engineers of automation to make the operational limits transparent (Seppelt and Lee, 2007).

A crucial part of ensuring system transparency is to ensure that Common Ground (CG) has been established. CG is defined as the sum of two or more persons' (or agents') mutual beliefs, knowledge and suppositions (Clark, 1996; Heath et al., 2002; Hoc, 2001; Huber and Lewis, 2010; Keysar et al., 1998; Stalnaker, 2002; Vanderhaegen et al., 2006). CG may be achieved by ensuring that the driver receives feedback that acknowledges that inputs have been registered or that an error in the input transmission has occurred. According to Brennan (1998), feedback of this type is of utmost importance in achieving CG. Ensuring that CG is achieved is crucial in a highly automated vehicle as the driving task is distributed between driver and automation, and to succeed, both entities need to be aware of the other entities' actions

(Hollan et al., 2000; Hutchins, 1995a, 1995b; Wilson et al., 2007). An example of how this is applied in human–human communication is the use of acknowledging phrases, such as 'roger', when acknowledging statements in nuclear power plant control rooms and on the flight deck (Min et al., 2004).

Furthermore, if the system provides continuous, timely and task-relevant feedback to the driver during, for example, highly automated driving, it is possible to reduce the cognitive effort of understanding the system state and whether user inputs are registered or not when it is time to resume manual control (Brennan, 1998; Clark and Wilkesgibbs, 1986; Sperber and Wilson, 1986). According to Patterson and Woods (2001), the purpose of the handover is to make sure that the incoming entity does not have an incorrect model of the process state, is aware of significant changes to the process, is prepared to deal with effects from previous events, is able to anticipate future events, and has the necessary knowledge to perform their duties. This is supported by research from Beller et al. (2013) who found that drivers who received automation reliability feedback were on average 1.1 seconds faster in responding to a failure, which, according to Summala (2000), is approximately the time it takes to respond to an unexpected braking event during manual driving.

Evidently, appropriate feedback may reduce the time needed for a successful take-over as it could allow the driver to anticipate the need to intervene. Research by Kircher et al. (2014) has shown that drivers adapt their usage of automation by disengaging DA systems before operational limits are reached.

These insights indicate that drivers are able to anticipate when to disengage automation in contemporary systems to ensure safe transfers of control. This does not necessarily mean that drivers will be able to adapt in such a way using systems in the future, as the majority of the driving task will be automated, and the driver will be less involved in the driving task.

2.2 PRINCIPLES OF COMMUNICATION

In order to demonstrate the importance of communication and feedback, Norman (1990) posited a thought experiment. In the first part of the experiment, an airline pilot handed over control to the aircraft autopilot. In the second part of the thought experiment, control was handed to a co-pilot instead of the autopilot. Norman argued that the task is 'automated' from the captain's point of view in both examples. If an event were to occur mid-flight to create an imbalance in the aircraft, both autopilot and co-pilot have the ability to successfully compensate for the imbalance. However, there is a large difference in the way that the information about compensatory actions would be communicated to the captain. In the case of the autopilot, the compensatory behaviour would only be communicated through the changes of controller settings in the cockpit and could easily be missed by the crew, as they are out-of-the-loop. In the case of the co-pilot, compensatory actions would be executed by means of the physical movements of the co-pilot that are required to change controller settings and to move control yokes as well as by verbal communication, such as stating, 'The aircraft started to bank to the left so I have had to increase the right wing down setting of the control wheel'. Thus, in the case of the co-pilot, the compensatory actions taken would be significantly more obvious to the captain. Examples of such situations are given in Section 2.3.

In a DA context, a similar but strictly theoretical scenario could be that the DA system compensates for an imbalance in the steering system caused, for example, by a partially deflated tire, by counter-steering. If the vehicle utilised steer-by-wire technology, with which the physical connection between the wheels and the steering wheel is replaced by sensors, torque motors and servos (e.g. Nexteer, 2017), it would be possible for the DA system to compensate for this imbalance by adjusting the position of the wheels to produce a counter-steering effect without moving the steering wheel. If this was the case, and the driver was prompted to resume control, it is very unlikely that the transient manoeuvre would be carried out in a safe manner as the vehicle would suddenly turn as the counter-steering ceased at the moment of transfer of control.

If the DA system was to mimic the counter-steering effect of the steering wheel on the wheels, thus quite clearly communicating that compensatory action is being taken, the transfer of control may be successful. The driver is then able to continue applying counterforce by maintaining the same steering wheel angle as the DA system when the compensatory action was carried out. Contemporary research has found similar results, whereby the steering torque has changed from when drivers relinquish control to when they resume control from the automation; such a change in torque is experienced when vehicle velocity changes during transit (Russell et al., 2016).

Norman concluded that when automated systems work in silence, not communicating actions and state, and with the operators being out-of-the-loop, a sudden automation failure or shutdown due to reaching the operational limits of the systems might take the operators by surprise, leading to an unrecoverable situation. Furthermore, he stated that feedback is essential in updating operators' mental models of the system (thus achieving CG) and that feedback is essential for appropriate monitoring and error detection and correction (Norman, 1990). In an effort to reduce the gulf of evaluation (Norman, 2013) and to increase behavioural transparency of automation to a similar level to that of a human co-pilot, the author would like to introduce a 'Chatty Co-Driver paradigm' (Eriksson and Stanton, 2016; Hoc et al., 2009; Stanton, 2015). It is argued that human–automation interaction may be facilitated by the application of the cooperative principle (Grice, 1975) as a heuristic in designing the automation human–machine interface and feedback.

To ensure that system feedback is communicated in an effective manner, it has to adhere to certain principles. As part of the cooperative principle, Grice (1975) posited four *maxims* for successful conversation: Quantity, Quality, Relation and Manner detailed as follows:

1. The *Maxim of Quantity (MoQu)* states that contributions should be made as informative as required without contributing more information than required to do so.
2. *The Maxim of Quality (MoQa)* states that information provided should not be false and that it should be supported by adequate evidence.
3. *The Maxim of Relation (MoR)* states that the information contributed should be relevant to the task/activity, context and need in the current situation.

4. *The Maxim of Manner (MoM)* relates to how the information is provided rather than what information is provided. The maxim states that obscurity and ambiguity should be avoided, and that information should be conveyed briefly and in an orderly manner.

The Gricean maxims were chosen as the underlying framework for assessing human–automation interaction, as the Gricean maxims have been considered seminal in the development of pragmatics and are very influential in the areas of linguistic semantics, philosophy of language, philosophy of mind and psychology (Clark, 2013, pp. 63–34). Moreover, the Gricean maxims have been used to assess human–human interaction (Attardo, 1993; Rundquist, 1992; Surian et al., 1996), and one study used the Gricean maxims to design automated system etiquette in aviation resulting in an increase in operator performance when automation etiquette was optimal and vice versa (Sheridan and Parasuraman, 2016). It is therefore argued that it is fitting to apply the fundamental ideas of human–human communication in this novel way, which examines human–automation communication allowing a concept for human–automation communication to naturally arise rather than applying more refined models targeted at more specific communication theories devised for the application in other areas than automation.

The potential use of the Gricean maxims in assessing information quality and its effects on coordination in human–automation interaction has yet to be explored. In DA, the crucial part of the coordinative act is the transfer of vehicle control between man and machine. A successful control transition necessitates a meaningful exchange of information between driver and automation. By applying the Chatty Co-Driver paradigm, feedback could be designed in a similar fashion to the feedback provided by a human co-pilot in aviation – for example, by verbally announcing decisions and actions – and changes to controller inputs are made salient through the physical movements of the controllers by the human pilot. By applying the Gricean maxims in this way, it might be possible to determine when and how to display information in a contextually and temporally relevant way.

2.3 CASE STUDIES

Contemporary DA is still in its infancy, and as a result its availability for study is limited or restricted to test tracks, pre-determined test-beds and manufacturer prototypes. Therefore, the perspectives from Section 2.2 will be applied to the aviation domain in which autopilots are commonplace and where human–automation interaction is prevalent.

2.3.1 METHOD

To illustrate the explanatory value of the maxims, they will be applied to two case studies, the Air France 447 crash (Bureau d'Enquêtes et d'Analyses pour la sécurité de l'aviation civile, 2012, BEA) and the ComAir Flight 3272 crash (National Transportation Safety Board, 1997, NTSB), as aviation is a domain where communication breakdowns are regarded as a serious threat to safety (Molesworth and

Estival, 2015). The analysis of the case studies is based on data from the official investigations by the BEA (2012) and NTSB (1997), as well as previous analyses by Eriksson and Stanton (2015). These incidents were chosen because the official investigations deemed poor feedback and communication as amongst the primary contributing causes of the accidents.

2.3.2 AIR FRANCE 447

2.3.2.1 Synopsis

On 31 May 2009, an Airbus A330-200 operated by Air France was scheduled to carry 216 passengers and 12 crew members on Flight AF447 between Rio de Janeiro Galeão and Paris Charles de Gaulle. The captain was assigned as Pilot Not Flying (PNF) and one of the co-pilots was assigned as Pilot Flying (PF). AF447 was cruising at flight level 350 (FL350) in calm conditions at the start of the Cockpit Voice Recorder (CVR) recording, just after midnight. At 01:52, the captain woke the resting co-pilot and requested he take his place as the PNF; 8 minutes later the PF briefed the newly arrived co-pilot. In the briefing, the PNF mentioned that the recommended maximum altitude (REC MAX) was limiting their ability to climb above a turbulent area due to a higher than expected temperature. Following the briefing, the captain left the PF and the replacement PNF to continue the flight. At 02:08, the PNF suggested a heading alteration of 12° degrees to avoid the turbulent area; the crew also decided to decrease speed from Mach 0.82 to Mach 0.8 (~529 kt) and to turn on engine de-icing. At 02:10:05 the autopilot and the auto-thrust disconnected, likely due to all the pitot probes of the Airbus A330-200 freezing over, resulting in unreliable speed readings. Following the autopilot disconnection, the PF said, 'I have the controls', indicating he was in control of the aircraft. The PF simultaneously gave a nose up and left input as a response to the aircraft rolling to the right at the time of autopilot disconnecting. The actions of the PF triggered a stall warning as the angle of attack increased beyond the flight envelope boundaries at Mach 0.8 (Figure 2.1, Angle of Attack, AOA > 4°). This was the start of a series of events of miscommunication between the flight crew and between the crew and aircraft, as demonstrated in Table 2.1.

During the remaining 4 minutes and 23.8 seconds, the aircraft continued to climb, leaving the lift envelope, trading kinetic energy for potential energy, until it unavoidably started to descend. The PF, unaware of the situation, continued to apply nose-up inputs which further increased the AOA which, from 02:12 to the end of the flight,

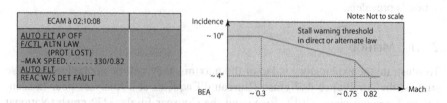

FIGURE 2.1   (a) Information readily available for PNF at the time of autopilot disconnect. (b) AOA threshold for stall warning at different speeds. *Source*: BEA www.bea.aero.

TABLE 2.1

Voice Data Recording Transcript and Flight Data Recording Coupled with an Analysis of the Human–Machine and Human–Human Communication from a Perspective of the Individual Gricean maxims

Time	Voice Data Recording		Flight Data Recording	MoQu	MoQa	MoR	MoM
2 h 10 m	**PNF**	**PF**					
04.6s		Do you us to put it on ignition start	Cavalry charge	✓			✓
05s			Autopilot disconnection warning				
			Autopilot disconnects				
			Roll angle changes from 0° to 8.4° in 2 seconds, sidestick neutral				
			Pitch attitude is 0°				
06s			Flight control law changes to alternate	✓	×	✓	×
06.4s		I have the controls		✓	✓	✓	✓
07s			PF sidestick positioned at 75% nose up				
			Pitch attitude increases to 11°				
			Vertical speed increase to 5200 ft/min				
07.5s	Alright		Flight Director 1 & 2 becomes unavailable	✓	×	✓	✓
08s			Auto thrust is disengaged	×	×	×	×
			CAS changes from 274 to 156 kt				
09s			CAS is 52 kt	×	✓	×	×
09.3s		Ignition start					
10s			Stall warning is triggered	✓	✓	✓	✓
			AOA 1, 2 & 3 is 2.1°, 4.9°, 5.3°				

(Continued)

TABLE 2.1 (CONTINUED)

Voice Data Recording Transcript and Flight Data Recording Coupled with an Analysis of the Human–Machine and Human–Human Communication from a Perspective of the Individual Gricean maxims

Time	Voice Data Recording		Flight Data Recording	MoQu	MoQa	MoR	MoM
2 h 10 m	PNF	PF					
11.3s	What is that?						
12s			CAS is 55 kt	×	×	×	×
13.5s			Stall warning seizes	×	×	×	×
14s		We haven't got a good…		✓	✓	✓	✓
15.1s		We haven't got a good display…		✓	✓	✓	✓
15.9s	We've lost the the speeds so… engine thrust A T H R engine level thrust	…of speed					
17s			Flight director 1 & 2 becomes available again; active mode is HDG/ALT CRZ; CAS is 80 kt	×	×	×	×

Key: ✓, maxim fulfilled; ×, maxim violated.

Note: Voice data recording transcript and flight data recording temporally aligned between 2 h 10 min 04 s to 2 h 10 min 17 s into the flight.

were on average around 40°. The last recorded vertical speed value was −10,912 ft/min as the aircraft crashed into the Atlantic Ocean.

2.3.2.2 Analysis

At 02:10:04 a cavalry charge sounded in the cockpit of AF447, indicating that the autopilot had disengaged. An immediate change in roll angle from 0°–8.4° without any sidestick input followed the autopilot disengagement (Table 2.1). The PF responded appropriately, acknowledging the event by stating he had the controls, thereby making the PNF aware that manual control was resumed. At this point, the PF's task was to maintain control of the aircraft and the PNF's task was to identify the fault and ensure that the designated flight path was followed. The PNF does this by checking the instruments and the Electronic Centralised Aircraft Monitor (ECAM) display (Figure 2.1).

The ECAM is designed to provide pilots with information in a quick and effective manner and by displaying the corrective actions needed to resolve any errors. As the PNF tried to identify the reason for the unexpected disconnection of the autopilot by checking the ECAM messages (see Figure 2.1), there was nothing to indicate that the autopilot disconnect had anything to do with the pitot probes freezing over, causing inaccurate speed readings. The only ECAM indication of a speed-related error was a message that indicated that the maximum speed was Mach 0.82, which, according to the BEA investigation, could be misinterpreted as the aircraft being in an overspeeding situation.

At the time of the transition of control of the aircraft from autopilot to pilot, there was nothing to indicate why the transfer of control had occurred and what future actions needed to be taken to ensure the continued safe operation of the flight.

According to Patterson and Woods (2001), the main purpose of a transfer of control, or handover, is to ensure that the agent taking over control has a correct mental model of the current state of the process and is aware of any changes to the process. The agent resuming control must also be prepared to deal with the effects of previous events and needs to be able to anticipate future events. These requirements were not fulfilled at the time of the transfer of control, as neither the PF nor the PNF succeeded in identifying the underlying cause of the autopilot disengagement in the initial period after control was transferred. As the PF and PNF failed to create an accurate mental model of the state of the current system due to a lack of system feedback, their responses to unfolding events were inappropriate for the situation, and thus resulted in worsening, rather than improving, the situation.

It was possible to identify several violations of the Gricean maxims in the moments where control was transferred from autopilot to PF, as shown in Table 2.1. The absence of any information related to the pitot readings being inaccurate is a clear violation of the MoQu, as information clearly was not sufficient to provide the PF/PNF with the necessary information to assess the situation.

The MoQa was also violated as the MAX SPEED information was of no help in resolving the situation, as at the time of the incident, the aircraft was nowhere near an over-speeding situation. Thus, information indicating an over-speeding risk was provided although no underlying evidence of such a risk was present. Furthermore, the MoR was violated as the MAX SPEED information was irrelevant to the context at hand and did not assist in creating an accurate mental model. The MoM was also violated by the ambiguous nature of the speed warning which indicated a risk of over-speeding rather than the erroneous readings of the pilot probes.

2.3.3   ComAir 3272

2.3.3.1   Synopsis

Comair Flight 3272, an Embraer EMB 120 Brasilia, departed Covington, Kentucky, on its way to the Detroit Metropolitan/Wayne Country Airport on 9 January 1997 at 15:08, with two crew members, one flight attendant, and 26 passengers on board. It was a routine flight with some intermittent, light chop. At 15:52.13, air traffic control (ATC) cleared the pilots to descend to 4000 feet mean sea level (ft msl). The pilots acknowledged and complied with the clearance. At 15:54.05, the pilot initiated a left turn of 090° by changing the heading setting for the autopilot. According to Flight Data Recorder (FDR) data, at 15:54.08 the aircraft was in a shallow left bank remaining at 4000 ft msl. As the aircraft reached its assigned altitude of 4000 ft msl, the autopilot switched mode from 'Altitude Pre-Select (arm)' to 'Altitude Hold'. At 15:54.10, the aircraft was flying at an airspeed of 156 kt and the roll attitude of the aircraft had steepened to an approximate 23° left wing down bank, causing the autopilot to turn the control wheel position (CWP) (CWP is the control wheel that controls the aircraft roll angle) in a right wing down (RWD) direction. Unfortunately, this corrective action had no effect, as the left wing down angle continued to steepen.

At 15:54.15, the left bank continued to steepen whilst the autopilot continued to apply compensatory measures by turning the CWP to apply right wing down commands. Approximately 15 seconds later the bank angle was steepening beyond 45°, at which point the autopilot disconnected and the stick-shaker engaged. In less than 2 seconds after the disconnection of the autopilot, the CWP moved from 18° right wing down command to 19° left wing down command, the roll increased from 45° left to 140° and pitch attitude decreased from 2° nose up to 17° nose down. At 15:54.25, the sound of the stick-shaker ceased, followed by an utterance of 'Oh' from both the captain and the first officer.

Approximately 30 seconds after the autopilot disengagement, the aircraft's left roll attitude increased beyond 140° and the pitch attitude approached 50° nose down. According to the CVR data, the ground proximity warning system triggered the 'bank angle' aural warning follow by three chimes and the autopilot aural warning repeatedly until the end of the recording.

2.3.3.2   Analysis

As Comair 3272 descended to 4000ft msl, the pilots were instructed to carry out a heading change to 090° by entering the new heading into the autopilot. The autopilot responded by banking to the left to achieve the desired heading by adjusting the ailerons and thus, the CWP in the cockpit. During the descent, the aircraft had accumulated ice on the aerofoil, and, as the aircraft entered the left bank, NTSB theorise that the change in bank angle caused asymmetrical ice self-shedding which combined with aileron deflection effects causing the left bank angle to steepen. As the bank angle steepened beyond the instructed bank angle, the autopilot's design logic responded by instructing the aileron servos to move the ailerons and the CWP in a RWD fashion to counter the increasing roll rate. When the autopilot was working in the heading mode, the maximum commanded bank angle was 25°, which was exceeded as the bank angle increased; this limitation reduced the compensatory behaviour that the autopilot could carry out. When the maximum autopilot bank angle of 25° is exceeded, no alarm nor any other cues are issued in the cockpit; the first indication occurs when the bank angle exceeds 45° and the autopilot automatically disengages. As the bank angle steepened beyond 45°, the autopilot disengaged and the following events occurred simultaneously: The stick-shaker activated. A repeated 'ding, ding, ding, autopilot' aural alert sounded, red autopilot fail/disengage lights were illuminated on the autopilot, flight control and master warning panels. The aircraft went from 45° left bank to 140° left bank, the CWP changed from 18° RWD to 19° left wing down (LWD) and the pitch attitude changed from 2° nose up to 19° nose down.

This particular accident is similar to the one described in Section 2.3.2. The official investigation suggested that if one of the pilots had been gripping the control wheel to monitor automation performance, the error would likely have been detected. However, as humans are notoriously bad at maintaining sustained attention for longer periods of time (Mackworth, 1948), it is arguable that such behaviours are unlikely and should not be expected (Heikoop et al., 2016; Molloy and Parasuraman, 1996).

From a Gricean perspective, the control wheel was designed in such a fashion that any changes in CWP were not salient enough. The control wheel design does provide information that is both correct and relevant to the task, and the design carries enough information to understand what is going on, thus fulfilling the requirements of the MoQu, MoQa and MoR. However, the control wheel does not provide unambiguous and salient information, thereby violating the MoM.

The autopilot feedback was also found insufficient. As the aircraft passed the 25° command limit of the autopilot, no alarm sounded and no feedback was shown on the cockpit displays. No feedback was issued at all until the bank angle exceeded 45°, and the autopilot disconnected along with the autopilot warning. As there was no indication of the aircraft approaching and exceeding the autopilot's performance boundaries, the pilots were unaware of the severity of the aerodynamic degradation. From a Gricean perspective, the lack of warning violates the maxim of quantity, as there was no information presented to the pilots regarding the bank angle exceeding 25°. Furthermore, the lack of information implies that the systems were working normally, thus violating the MoQa and the MoR, as the system was past its interaction boundary (the bank angle envelope of 25° in which the autopilot could apply RWD/LWD commands), and had feedback been presented, the pilots would have had a chance to compensate manually. Due to the lack of any feedback, the MoM was violated in the sense that there is no information/feedback that could be assessed using the maxim.

The FDR data indicated that the pilots responded by applying RWD commands on the control wheel within 1 second after the autopilot disconnected to counter the sudden increase in left bank. If changes to the CWP had been more salient, clearly indicating both the corrective actions that were taken by the autopilot in terms of bank angle and also the force that was applied to the control wheel, the pilots would have had a clear indication of the actions taken by the autopilot. Furthermore, had there been a warning or a notice on one of the cockpit displays when the autopilot exceeded its interaction boundary, they might have been made aware of the autopilot struggling to maintain the bank angle required for the heading change due to the degrading aerodynamic characteristics of the aerofoil.

Such an indication would have been in accordance with the MoR, by providing temporally and contextually relevant information. This would have allowed the pilots to resume manual control earlier and would have allowed them to increase airspeed and abort the heading change, thus reducing the bank angle, before the autopilot reached its 45° performance boundary and disconnected.

2.3.4 Summary of Case Studies

To summarise the results of the case studies from a Gricean perspective, it is clear that several maxim violations contributed to the crashes of Air France AF447 and Comair 3272. The maxims highlighted issues relating to lack of, or non-salient, feedback to mode changes and system actions. Furthermore, the use of the maxims allowed for the identification of an issue related to the non-salient feedback of a piece of equipment that otherwise functioned according to specification, namely the autopilot-controlled changes to CWP as shown in Table 2.2.

The maxims show promise in identifying flawed communication in the human-cockpit system; therefore they have potential as a tool to inform the design of similar systems. Such a system could be highly automated driving systems in which interaction could be designed in such a way that events similar to those in the previously mentioned case studies may be avoided in future automated vehicles.

2.4   LESSONS LEARNT

The case studies in Section 2.3 have illustrated how operator mental models deteriorate as they are removed from the control loop. In the case studies, this resulted in the loss of life due to crashes when the operator suddenly had to resume control with an

---

TABLE 2.2
A Summary of the Maxim Violations in Each Part of the Case Studies

	AF447 Speed Readings	Comair Control Wheel	Comair Autopilot Warning
Maxim of Quantity	×	✓	×
Maxim of Quality	×	✓	×
Maxim of Relation	×	✓	×
Maxim of Manner	×	×	×

Key: ✓, maxim fulfilled; ×, maxim violated.

inaccurate mental model after prolonged exposure to automated systems with no, or poor, feedback. The motor industry has, until recently, been spared such issues, but with the introduction of new technology, drivers have been able to relinquish control of certain sub-tasks of driving, resulting in the fatal crash of a Tesla Model S in Florida in 2016 (Office of Defects Investigation, 2017). These systems have allowed drivers to take their feet off the pedals, and more recently, have their lane position controlled automatically as well. Although drivers have yet to be able to fully relinquish control to the vehicle, recent amendments to the Vienna Convention on Road Traffic now enable drivers to be fully hands- and feet-free as long as the system can be overridden or switched off by the driver (Miles, 2014).

This amendment will allow autonomous vehicles owned by the public to be driven hands- and feet-free on public roads. The change in legislation implies significant changes to the driving task. Whilst contemporary DA systems require intermittent driver feedback, by, for example, touching the steering wheel or the indicator stalk, thus maintaining some level of driver engagement (Naujoks et al., 2015), the amended section of the Vienna Convention on Road Traffic will allow the driver to be completely out-of-the-loop. As drivers are given more freedom to relinquish the driving task to the automated systems, their role in the DA system is altered, approaching that of a pilot flying by means of the autopilot (Hollnagel and Woods, 1983; Parasuraman et al., 2000; Woods, 1996). The driver will, much like a pilot, have to monitor the DA system for errors or deviant behaviours and will be expected to resume control when something goes wrong or when operational limits are reached (Hollnagel and Woods, 2005; SAE J3016, 2016; Stanton et al., 1997).

As shown in Section 2.3, pilots may fail to successfully recover their aircraft when the automated systems fail to cope, despite being subjected to extensive training before receiving their pilot's license as well as regular training sessions of fringe scenarios in simulators. This training is supposed to prepare pilots for rare occurrences, such as degraded aerodynamic properties of the aerofoil due to icing, autopilot operational limits, or the safety envelope of an Airbus A380 in different laws, which allows them to adapt to ever-changing conditions.

The fact that highly trained, professional pilots sometimes struggle to cope with such high-complexity situations is particularly worrying, as drivers soon will be able to purchase vehicles equipped with DA that enables similar features in the vehicle; this will allow hands- and feet-free driving without any additional training.

Section 2.3 showed that breakdowns in human–machine communication could be attributed to the cause of the accidents. In the aviation domain, communication breakdowns are considered a serious threat to safety (Molesworth and Estival, 2015). The same should be the case in the driving domain, especially since there is a differing degree of complexity in the DA domain caused by a large fleet of vehicles occupying a relatively small space where non-connected road users, such as pedestrians, cyclists and wild animals, co-exist. This puts even higher demands on the engineers responsible for designing the driver interface for the DA systems.

The case studies in Section 2.3 clearly identify, by means of the Gricean maxims, non-salient feedback, non-existing feedback and erroneous feedback as contributing factors in the accidents that transpired. Humans are known to be poor monitors in situations where passive monitoring is required (Heikoop et al., 2016; Mackworth, 1948; Molloy and Parasuraman, 1996); and as they are removed from the driving loop, measures must be taken to ensure that DA systems do not succumb to similar issues as described in the case studies. Aviation accidents may be rare, but the consequences of similar communication issues being present in DA systems, coupled with the exposure of untrained operators to such systems, has the potential to be dire.

To ensure that incidents caused by poor communication between man and machine are reduced, Human Factors Engineers must aim to reduce the gulf of evaluation, thereby retaining driver Common Ground and thus a representative mental model of the DA system (Klein et al., 2004; Norman, 1990). This could potentially be done using a 'Chatty Co-Driver paradigm' (Eriksson and Stanton, 2016; Hoc et al., 2009; Stanton, 2015). The Chatty Co-Driver paradigm tries to reduce the gulf of evaluation by providing feedback similar to that of a co-pilot (Wiener, 1989) thus increasing system transparency (Christoffersen and Woods, 2002). Such feedback could entail information regarding what the vehicle sensors are able to register, thus providing an indication of the limitations of the vehicle radars as well as whether the vehicle has picked up objects entering the trajectory of the automated vehicle (Jenkins et al., 2007; Stanton et al., 2011).

Furthermore, as shown by Beller et al. (2013), contextually relevant information could reduce the time of transfer of control as it brings the driver back into the control loop by restoring driver mental models, and it thus reduces the likelihood of an escalation of cognitive activities (Woods and Patterson, 2000). Research by Eriksson and Stanton (2015) indicates that information needs change depending on contextual and temporal factors. Based on the aforementioned maxims of successful conversation (Grice, 1975), the author proposes that by adhering to MoQu, MoQa, MoR and MoM in designing automated systems, the gulf of evaluation could be reduced and system transparency could be increased.

2.5   CONCLUDING REMARKS

This chapter highlighted some of the problems drivers will face when encountering DA systems, such as automation surprises and ending up out-of-the-loop. When drivers are out-of-the-loop and a change in operational conditions occur, requiring a control transition from DA to manual control, the driver must have access to sufficient information to safely complete the transition. Due to the lack of DA systems available to the public, two case studies were presented from the domain of aviation. It was demonstrated that highly trained professionals failed to safely resume control after a control transition due to poor or lacking feedback from the automated system.

By applying the Gricean maxims of successful conversation to the case studies from a human–automation interaction perspective, the author was able to identify lacking feedback in different components of the pilot interface. In addition, maxims could be used as a means to bridge the gulf of evaluation in contemporary DA systems, by allowing the DA system to act like a Chatty Co-Driver, thereby increasing system transparency and reducing the effects of being out-of-the-loop. Some high-level proposals for design based on the maxims are provided in Table 2.3.

TABLE 2.3

Design Proposals for Information Presentation Based on the Gricean Maxims

	Design Proposal
Maxim of Quantity	• Avoid oversaturating the information presentation modality by limiting the information presented to the bare necessity.
Maxim of Quality	• Ensure that the information presented is based on reliable data. • If the information is based on unreliable data, ensure that it is made clear in the human–machine interface.
Maxim of Relation	• Ensure that the presented information is contextually relevant, for example, based on Michon's (1985) operational, tactical and strategic levels of information. • Do not present information that is non-relevant to accomplish the task.
Maxim of Manner	• Avoid ambiguity by, for example, reducing similarities of auditory tones or increasing contrast between similar modes in the visual human–machine interface. • Ensure that the right information is presented in the right modality; for example, urgent signals could be presented both haptically, aurally and visually whereas non-urgent information could be presented visually in the human–machine interface (Meng and Spence, 2015).

2.6 FUTURE DIRECTIONS

The appropriate design of system feedback is a key component in the deployment of safe automated vehicle systems. Whilst the Gricean maxims show promise in shedding light on the design of such feedback, our current understanding of designing interfaces based on human–human communication is limited. Chapter 4 tries to identify the time span drivers require to safely resume control from an automated vehicle with the level of feedback available in contemporary vehicles. Establishing a baseline time span for the required time to resume control is crucial, as these findings can be further contrasted through the introduction of human–machine interface elements to facilitate the transition (Chapter 7) in accordance with the Gricean maxims. Indeed, Beller et al. (2013) found that drivers who received automation reliability feedback were on average 1.1 seconds faster to respond to a failure, indicating that the right type of feedback can reduce the required time to intervene.

REFERENCES

Andre, A., Degani, A. (1997). Do you know what mode you're in? An analysis of mode error in everyday things. In M. Mouloua and J. M. Koonce (Eds.), *Human-Automation Interaction: Research and Practice*. Mahwah, NJ: Lawrence Erlbaum, pp 19–28.

Attardo, S. (1993). Violation of conversational maxims and cooperation – The case of jokes. *Journal of Pragmatics*, vol 19, no 6, pp 537–558.

Bainbridge, L. (1983). Ironies of automation. *Automatica*, vol 19, no 6, pp 775–779.

Beiker, S. A. (2012). Legal aspects of autonomous driving: The need for a legal infrastructure that permits autonomous driving in public to maximize safety and consumer benefit. *Santa Clara Law Review*, vol 52, pp 1145–1561.

Beller, J., Heesen, M., Vollrath, M. (2013). Improving the driver-automation interaction: An approach using automation uncertainty. *Human Factors*, vol 55, no 6, pp 1130–1141.

Brennan. (1998). The grounding problem in conversations with and through computers. In S. R. F. R. J. Kreuz (Ed.), *Social and Cognitive Psychological Approaches to Interpersonal Communication*. Hillsdale, NJ: Lawrence Erlbaum, pp 201–225.

Bureau d'Enquêtes et d'Analyses pour la sécurité de l'aviation civile. (2012). *Final Report: On the Accident on 1st June 2009 to the Airbus A330-203 Registered F-GZCP Operated by Air France Flight AF 447 Rio De Janeiro – Paris*.

Christoffersen, K., Woods, D. D. (2002). How to make automated systems team players. *Advances in Human Performance and Cognitive Engineering Research*, vol 2, pp 1–12.

Clark, B. (2013). *Relevance Theory*. Cambridge: Cambridge University Press.

Clark, H. H. (1996). *Using Language*. Cambridge: Cambridge University Press.

Clark, H. H., Wilkesgibbs, D. (1986). Referring as a collaborative process. *Cognition*, vol 22, no 1, pp 1–39.

Cranor, L. F. (2008). A framework for reasoning about the human in the loop. Paper presented at the 1st Conference on Usability, Psychology, and Security, San Francisco, CA.

Degani, A., Shafto, M., Kirlik, A. (1995). Mode usage in automated cockpits: Some initial observations. *IFAC Proceedings Volumes*, vol 28, no 15, pp 345–351.

Desmond, P. A., Hancock, P. A., Monette, J. L. (1998). Fatigue and automation-induced impairments in simulated driving performance. *Human Performance, User Information, and Highway Design*, no 1628, pp 8–14.

Endsley, M. R. (1996). Automation and situation awareness. In R. Parasuramanand M. Mouloua (Eds.), *Automation and Human Performance: Theory and Applications*. Mahwah, NJ: Lawrence Erlbaum.

Endsley, M. R., Onal, E., Kaber, D. B. (1997). The impact of intermediate levels of automation on situation awareness and performance in dynamic control systems. Human Factors and Power Plants, 1997. Global Perspectives of Human Factors in Power Generation. In *Proceedings of the 1997 IEEE Sixth Conference*, Orlando, FL, 1997, pp 7/7–712.

Eriksson, A., Stanton, N. A. (2015). When communication breaks down or what was that? – The importance of communication for successful coordination in complex systems. Paper presented at the 6th International Conference on Applied Human Factors and Ergonomics, Las Vegas, NV.

Eriksson, A., Stanton, N. A. (2016). The chatty co-driver: A linguistics approach to human-automation-interaction. Paper presented at the IEHF2016, Daventry, UK.

Eriksson, A., Stanton, N. A. (2017). Takeover time in highly automated vehicles: Noncritical transitions to and from manual control. *Human Factors*, vol 59, no 4, pp 689–705.

Gasser, T. M., Arzt, C., Ayoubi, M., Bartels, A., Bürkle, L., Eier, J. et al. (2009). Legal consequences of an increase in vehicle automation.

Grice, H. P. (1975). Logic and conversation. In P. Cole and J. L. Morgan (Eds.), *Speech Acts*. New York: Academic Press, pp 41–58.

Heath, C., Svensson, M. S., Hindmarsh, J., Luff, P., vom Lehn, D. (2002). Configuring awareness. *Computer Supported Cooperative Work (CSCW)*, vol 11, no 3–4, pp 317–347.

Heikoop, D. D., de Winter, J. C. F., van Arem, B., Stanton, N. A. (2016). Psychological constructs in driving automation: a consensus model and critical comment on construct proliferation. *Theoretical Issues in Ergonomics Science*, vol 17, no 3, pp 284–303.

Hoc, J. H. (2001). Towards a cognitive approach to human-machine cooperation in dynamic situations. *International Journal of Human-Computer Studies*, vol 54, no 4, pp 509–540.

Hoc, J.-M., Young, M. S., Blosseville, J.-M. (2009). Cooperation between drivers and automation: Implications for safety. *Theoretical Issues in Ergonomics Science*, vol 10, no 2, pp 135–160.

Hollan, J., Hutchins, E., Kirch, D. (2000). Distributed cognition: A new foundation for human-computer interaction research. *ACM Transactions on Human-Computer Interaction: Special Issue on Human-Computer Interaction in the New Millennium*, vol 7, no 2, pp 174–196.

Hollnagel, E., Woods, D. D. (1983). Cognitive systems engineering: New wine in new bottles. *International Journal of Man-Machine Studies*, vol 18, no 6, pp 583–600.

Hollnagel, E., Woods, D. D. (2005). *Joint Cognitive Systems Foundations of Cognitive Systems Engineering*. Boca Raton, FL: CRC Press.

Huber, G. P., Lewis, K. (2010). Cross-understanding: Implications for group cognition and performance. *Academy of Management Review*, vol 35, no 1, pp 6–26.

Hutchins, E. (1995a). *Cognition in the Wild*: Cambridge, MA: MIT Press.

Hutchins, E. (1995b). How a cockpit remembers its speeds. *Cognitive Science*, vol 19, no 3, pp 265–288.

Inagaki, T. (2003). Adaptive automation: Sharing and trading of control. In Hollnagel E. (Ed.), *Handbook of Cognitive Task Design*. Mahwah, NJ: Lawrence Erlbaum Associates, pp 147–169.

Jenkins, D. P., Stanton, N. A., Walker, G. H., Young, M. S. (2007). A new approach to designing lateral collision warning systems. *International Journal of Vehicle Design*, vol 45, no 3, pp 379–396.

Kaber, D. B., Endsley, M. R. (1997). Out-of-the-loop performance problems and the use of intermediate levels of automation for improved control system functioning and safety. *Process Safety Progress*, vol 16, no 3, pp 126–131.

Kaber, D. B., Endsley, M. R. (2004). The effects of level of automation and adaptive automation on human performance, situation awareness and workload in a dynamic control task. *Theoretical Issues in Ergonomics Science*, vol 5, no 2, pp 113–153.

Kaber, D. B., Riley, J. M., Kheng-Wooi, T., Endsley, M. R. (2001). On the design of adaptive automation for complex systems. *Cognitive Ergonomics*, vol 5, no 1, pp 37–57.

Keysar, B., Barr, D. J., Balin, J. A., Paek, T. S. (1998). Definite reference and mutual knowledge: Process models of common ground in comprehension. *Journal of Memory and Language*, vol 39, no 1, pp 1–20.

Kircher, K., Larsson, A., Hultgren, J. A. (2014). Tactical driving behavior with different levels of automation. *IEEE Transactions on Intelligent Transportation Systems*, vol 15, no 1, pp 158–167.

Klein, G., Woods, D. D., Bradshaw, J. M., Hoffman, R. R., Feltovich, P. J. (2004). Ten challenges for making automation a 'team player' in joint human-agent activity. *IEEE Intelligent Systems*, vol 19, no 6, pp 91–95.

Leveson, N. (2004). A new accident model for engineering safer systems. *Safety Science*, vol 42, no 4, pp 237–270.

Mackworth, N. H. (1948). The breakdown of vigilance during prolonged visual search. *Quarterly Journal of Experimental Psychology*, vol 1, no 1, pp 6–21.

Meng, F., Spence, C. (2015). Tactile warning signals for in-vehicle systems. Accid Anal Prev, 75, pp 333–346.

Merat, N., Jamson, A. H., Lai, F. C. H., Daly, M., Carsten, O. M. J. (2014). Transition to manual: Driver behaviour when resuming control from a highly automated vehicle. *Transportation Research Part F-Traffic Psychology and Behaviour*, vol 27, pp 274–282.

Michon, J. A. (1985). A critical view of driver behavior models: What do we know, what should we do? In L. Evans and R. C. Schwing (Eds.), *Human Behavior and Traffic Safety* (pp 485–524). New York: Plenum Press.

Miles, T. (2014, May 19). Cars could drive themselves sooner than expected after European push. Retrieved from https://www.reuters.com/article/us-daimler-autonomous-driving/cars-could-drive-themselves-sooner-than-expected-after-european-push-idUSKBN-0DZ0UV20140519.

Min, D., Chung, Y. H., Kim, B. (2001). An evaluation of computerized procedure system in nuclear power plant. In *Proceedings of the 8th IFAC/IFIP/IFORS/IEA Symposium on Analysis, Design, and Evaluation of Human-Machine Systems*, Kassel, Germany: IFAC, pp 597–602.

Molesworth, B. R. C., Estival, D. (2015). Miscommunication in general aviation: The influence of external factors on communication errors. *Safety Science*, vol 73, no 0, pp 73–79.

Molloy, R., Parasuraman, R. (1996). Monitoring an automated system for a single failure: Vigilance and task complexity effects. *Human Factors*, vol 38, no 2, pp 311–322.

National Transportation Safety Board. (1997). *In-flight icing encounter and uncontrolled collision with terrain Comair Flight 3272 Embraer EMB-120RT, N265CA*.

Naujoks, F., Purucker, C., Neukum, A., Wolter, S., Steiger, R. (2015). Controllability of partially automated driving functions – Does it matter whether drivers are allowed to take their hands off the steering wheel? *Transportation Research Part F: Traffic Psychology and Behaviour*, vol 35, pp 185–198.

Nexteer. (2017). NEXTEER automotive provides advanced cyber security for steering systems. Retrieved from https://www.nexteer.com/release/nexteer-automotive-provides-advanced-cyber-security-for-steering-systems/.

Norman, D. A. (1983). Design rules based on analyses of human error. *Communications of the ACM*, vol 26, no 4, pp 254–258.

Norman, D. A. (1990). The 'problem' with automation: Inappropriate feedback and interaction, not 'over-automation'. *Philosophical Transactions of the Royal Society B: Biological Sciences*, vol 327, no 1241, pp 585–593.

Norman, D. A. (2009). *The Design of Future Things*. New York: Basic Books.

Norman, D. A. (2013). *The Design of Everyday Things*. New York: Basic Books.

Office of Defects Investigation. (2017). Automatic vehicle control systems. *National Highway Traffic Safety Administration*.

Parasuraman, R. (2000). Designing automation for human use: Empirical studies and quantitative models. *Ergonomics*, vol 43, no 7, pp 931–951.

Parasuraman, R., Sheridan, T. B., Wickens, C. D. (2000). A model for types and levels of human interaction with automation. *IEEE Transactions on Systems, Man, and Cybernetics. Part A, Systems and Humans*, vol 30, no 3, pp 286–297.

Patterson, E. S. Woods, D. D. (2001). Shift changes, updates, and the on-call architecture in space shuttle mission control. *Computer-Supported Cooperative Work*, vol 10, no 3–4, pp 317–346.

Rankin, A., Woltjer, R., Field, J., Woods, D. D. (2013). "Staying ahead of the aircraft" and managing surprise in modern airliners. Paper presented at the 5th Resilience Engineering Symposium: Man aging trade-offs, Soesterberg, the Netherlands.

Rundquist, S. (1992). Indirectness: A gender study of flouting Grice's maxims. *Journal of Pragmatics*, vol 18, no 5, pp 431–449.

Rushby, J., Crow, J., Palmer, E. (1999). An automated method to detect potential mode confusions. Paper presented at the Digital Avionics Systems Conference, St. Louis, MO.

Russell, H. E. B., Harbott, L. K., Nisky, I., Pan, S., Okamura, A. M., Gerdes, J. C. (2016). Motor learning affects car-to-driver handover in automated vehicles. *Science Robotics*, vol 1, no 1.

SAE J3016. (2016). Taxonomy and definitions for terms related to driving automation systems for on-road motor vehicles, J3016_201609: SAE International.

Sarter, N. B., Woods, D. D. (1995). How in the world did we ever get into that mode? Mode error and awareness in supervisory control. *Human Factors*, vol 37, no 1, pp 5–19.

Sarter, N. B., Woods, D. D., Billings, C. E. (1997). Automation surprises. In G. Salvendy (Ed.), *Handbook of Human Factors and Ergonomics* (2nd edn). New York: Wiley.

Seppelt, B. D., Lee, J. D. (2007). Making adaptive cruise control (ACC) limits visible. *International Journal of Human-Computer Studies*, vol 65, no 3, pp 192–205.

Sheridan, T. B. (1995). Human centered automation: Oxymoron or common sense? *1995 IEEE International Conference on Systems, Man and Cybernetics*, Vols 1–5, Vancouver, BC, pp 823–828.

Sheridan, T. B., Parasuraman, R. (2016). Human-automation interaction. *Reviews of Human Factors and Ergonomics*, vol 1, no 1, pp 89–129.

Sperber, D., Wilson, D. (1986). *Relevance: Communication and cognition.* Oxford: Blackwell.

Stalnaker, R. (2002). Common ground (Speaker presupposition). *Linguistics and Philosophy*, vol 25, no 5–6, pp 701–721.

Stanton, N. A. (2015, March). Responses to autonomous vehicles. *Ingenia*, 9.

Stanton, N. A., Dunoyer, A., Leatherland, A. (2011). Detection of new in-path targets by drivers using Stop & Go Adaptive Cruise Control. *Applied Ergonomics*, vol 42, no 4, pp 592–601.

Stanton, N. A., Young, M. S., McCaulder, B. (1997). Drive-by-wire: The case of mental workload and the ability of the driver to reclaim control. *Safety Science*, vol 27, no 2–3, pp 149–159.

Strand, N., Nilsson, J., Karlsson, I. C. M., Nilsson, L. (2014). Semi-automated versus highly automated driving in critical situations caused by automation failures. *Transportation Research Part F-Traffic Psychology and Behaviour*, vol 27, pp 218–228.

Summala, H. (2000). Brake reaction times and driver behavior analysis. *Transportation Human Factors*, vol 2, no 3, pp 217–226.

Surian, L., Baron-Cohen, S., Van der Lely, H. (1996). Are children with autism deaf to gricean maxims? *Cognitive Neuropsychiatry*, vol 1, no 1, pp 55–72.

Swaroop, D., Rajagopal, K. R. (2001). A review of constant time headway policy for automatic vehicle following. Paper presented at the IEEE Intelligent Transportation Systems Conference Proceedings, Oakland, CA.

Vanderhaegen, F., Chalmé, S., Anceaux, F., Millot, P. (2006). Principles of cooperation and competition: Application to car driver behavior analysis. *Cognition, Technology and Work*, vol 8, no 3, pp 183–192.

Weick, K. E., Sutcliffe, K. M., Obstfeld, D. (2005). Organizing and the process of sensemaking. *Organization Science*, vol 16, no 4, pp 409–421.

Wiener, E. L. (1989). Human factors of advanced technology ("Glass Cockpit") transport aircraft. *NASA Contractor Report 177528*.

Willemsen, D., Stuiver, A., and Hogema, J. (2015). Automated driving functions giving control back to the driver: A simulator study on driver state dependent strategies. Paper presented at the 24th International Technical Conference on the Enhanced Safety of Vehicles (ESV).

Wilson, S., Galliers, J., Fone, J. (2007). Cognitive artifacts in support of medical shift handover: An in use, in situ evaluation. *International Journal of Human-Computer Interaction*, vol 22, no 1–2, pp 59–80.

Wolterink, W. K., Heijenk, G., Karagiannis, G. (2011). Automated merging in a Cooperative Adaptive Cruise Control (CACC) system. Paper presented at the Fifth ERCIM Workshop on eMobility, Vilanova i la Geltrú, Catalonia, Spain.

Woods, D. D. (1993). Price of flexibility in intelligent interfaces. *Knowledge-Based Systems*, vol 6, no 4, pp 189–196.

Woods, D. D. (1996). Decomposing automation: Apparent simplicity, real complexity. In R. Parasuraman, M. Mouloua (Eds.), *Automation and Human Performance: Theory and Applications*. Mahwah, NJ: Erlbaum.

Woods, D. D. and Patterson, E. S. (2000). How unexpected events produce an escalation of cognitive and coordinative demands. In P. A. Hancock and P. Desmond (Eds.), *Stress Workload and Fatigue*. Hillsdale, NJ: Lawrence Erlbaum Associates, pp 290–305.

Young, M. S., Stanton, N. A. (1997). Automobile automation: Investigating the impact on driver mental workload. *International Journal of Cognitive Ergonomics*, vol 1, no 4, pp 325–336.

Young, M. S., Stanton, N. A. (2002). Malleable attentional resources theory: A new explanation for the effects of mental underload on performance. *Hum Factors*, vol 44, no 3, pp 365–375.

Young, M. S., Stanton, N. A. (2007a). Back to the future: brake reaction times for manual and automated vehicles. *Ergonomics*, vol 50, no 1, pp 46–58.

Young, M. S., Stanton, N. A. (2007b). What's skill got to do with it? Vehicle automation and driver mental workload. *Ergonomics*, vol 50, no 8, pp 1324–1339.

Young, M. S., Stanton, N. A., Harris, D. (2007). Driving automation: Learning from aviation about design philosophies. *International Journal of Vehicle Design*, vol 45, no 3, pp 323–338.

3 A Toolbox for Automated Driving Research in the Simulator

The topic of automated driving is receiving an increasing level of attention from Human Factors researchers (Eriksson and Stanton, 2017b; Kyriakidis et al., 2017). Until recently, automated driving technology required intermittent driver feedback, for example, by touching the steering wheel, thus maintaining a level of driver engagement similar to manual driving (Naujoks et al., 2015). However, recent amendments to the Vienna Convention on Road Traffic now enable drivers to be fully hands- and feet-free as long as the system can be overridden or switched off by the driver (Miles, 2014). This amendment allows drivers to be 'out-of-the-loop' for prolonged periods of time, yet drivers are still expected to resume control when the operational limits of the automated driving system are approached (SAE J3016, 2016).

The availability of these highly automated driving systems may fundamentally alter the driving task (Hollnagel and Woods, 1983) and could give rise to 'ironies' and 'surprises' of automation similar to those proposed by Bainbridge (1983) and Sarter et al. (1997) in the context of process control and aviation. Indeed, several empirical studies have shown that drivers of highly automated cars often respond slowly when manual intervention is necessary (De Winter et al., 2014; Jamson et al., 2013; Stanton and Young, 1998; Stanton et al., 1997; Young and Stanton, 2007). In light of this, intermediate forms of automation have been deemed hazardous as drivers are required to be able to regain control at all times (Casner et al., 2016; Seppelt and Victor, 2016; Stanton, 2015). In order to study these psychological phenomena and develop effective human–machine interfaces for supporting drivers of future automated cars, the driving simulator is seen as a viable option (Boer et al., 2015; Eriksson et al., 2017).

3.1 SIMULATORS

Driving simulators have been used since the beginning of the 1930s (Greenshields, 1936) and Human Factors research into automated driving has been ongoing since the mid-1990s (Nilsson, 1995; Stanton and Marsden, 1996). As the motor industry advances towards highly automated driving, research conducted in driving simulators becomes increasingly important (Boer et al., 2015). Compared with on-road testing, driving simulators allow drivers' reactions to new technology to be measured in a virtual environment, without physical risk (Carsten and Jamson, 2011; De Winter et al., 2012; Flach et al., 2008; Nilsson, 1993; Stanton et al., 2001; Underwood et al., 2011).

Furthermore, driving simulators offer a high degree of controllability and reproducibility, and they provide access to variables that are difficult to accurately determine in the real world (Godley et al., 2002), such as lane position and distance to roadway objects (Santos et al., 2005; van Winsum et al., 2000). Most driving simulators offer flexibility in designing bespoke plug-ins through application programming interfaces (APIs). With Open Source software efforts in driving simulation such as OpenDS (OpenDS, 2017), it is likely that the use of driving simulators will grow in the coming years.

3.1.1 STISIM

STISIM is a popular driving simulator that is used for research purposes (Large et al., 2017; Large et al., 2016; McIlroy et al., 2016; Neubauer et al., 2014, 2016; Park et al., 2015). The STISIM driving simulator software comes with an 'automated driving' feature accessible through the Scenario Definition Language (SDL) (Allen et al., 1999a; Allen et al., 1999b; Allen et al., 2003; Allen et al., 2001). SDL-based automation allows the researcher to enable or disable automated lateral and/or longitudinal control through the Control Vehicle (CV) event by specifying a distance down the road at which point the event should be triggered, and what mode change should occur (e.g. the script '2000, CV, speed, 2' initiates automated control of both steering and speed when the participant has travelled 2000 m along the road). The STISIM documentation states that their automated driving feature is intended for driver training (Allen et al., 1999b), an approach also taken by other driving simulator manufacturers (e.g. De Winter et al., 2007). Indeed, by enabling automated control of speed, the driver can fully concentrate on learning how to steer, or vice versa, by enabling automated control of steering the learner driver can concentrate on how to accelerate and stop the car. This type of automation is sufficient when it comes to research where the researcher does not want the driver to be able to (dis)engage the automation or change the automation modes. The CV event has been successfully used in this manner (Funke, 2007; Funke et al., 2016; Neubauer, 2011; Neubauer et al., 2012; Neubauer et al., 2016; Saxby et al., 2016; Saxby et al., 2008). However, if the aim of the research is to understand how drivers *interact* with automated driving systems, as for example in Kircher et al. (2014), Eriksson and Stanton (2017b) and Eriksson et al. (2017), this type of hard-coded automation is not sufficient. The ability to study interactive behaviour is of prime importance, as it allows researchers to capture tactical behaviours, such as when, where and how a task is carried out (Kircher et al., 2017).

With the goal of enabling human–automation interaction, bespoke software using the STISIM V3 Build 3.07.04 Open Module was developed in Visual Basic 6 (VB6). Although this implementation of automated driving is platform specific, the toolbox can be easily implemented on other platforms that offer an API or software development kit (SDK) by translating the subroutines into the programming language supported by the simulator in question, with the requirement that the lead-vehicle can be identified and queried for information such as speed.

Open Module is a plug-in feature of the STISIM platform that allows researchers to implement their own modules using unmanaged code (e.g. VB6 or C++). One of the functions of Open Module is the 'Update'-function, which is called once every simulation frame, just before the graphics display updates. The 'Update'-function allows the researcher to directly control the behaviour of a vehicle via pedal and steering input, a functionality that was utilised in developing my toolbox. My algorithm toolbox consists of several subroutines, each responsible for a part of the vehicle control, allowing lateral and longitudinal automation to be used separately or in conjunction. The functionality of the toolbox algorithms is detailed subsequently.

3.2 ALGORITHMS

In this section, we describe my algorithms consisting of two parts: (1) longitudinal control and (2) lateral control. This structure enables the simulation of different levels of automated driving, ranging from manual driving and adaptive cruise control (ACC) (i.e. automated longitudinal control) to highly automated driving (i.e. automated longitudinal and lateral control), as shown in Figure 3.1. The manual mode is void of any automated features, meaning that the operation of the vehicle is dependent on the human driver only.

The ACC mode controls the vehicle's velocity by providing control signals to the throttle and brake inceptors in order to drive the vehicle at a target velocity (Figure 3.2). The target velocity is set by the driver or is dictated by the velocity of a slower vehicle within sensor range. ACC is an integral part of achieving highly automated driving and is now commonly available in production vehicles. ACC on

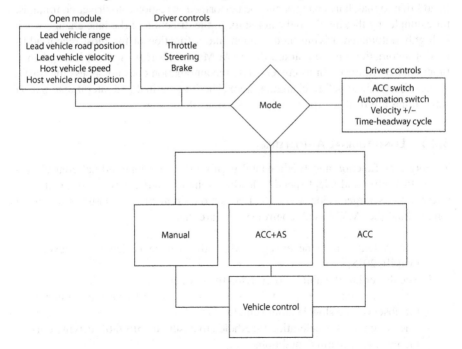

FIGURE 3.1 A functional block diagram of the automated driving toolbox. Figure 3.2 illustrates the functionality of the ACC subroutines.

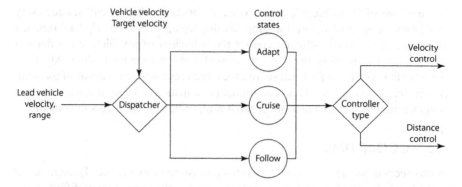

FIGURE 3.2   Functional block diagram of the longitudinal (ACC) control algorithm.

production vehicles utilises a radar unit attached to the front of the vehicle that keeps track of any leading vehicles, feeding the cruise control algorithm with the distance to the lead-vehicle which is used to compute the desired velocity to maintain the selected time-headway. The behaviour and rapidity of the manoeuvres varies with the control states, which are determined by the subroutine in Section 3.2

The highly automated driving mode incorporates the functionality of the ACC feature, with the addition of automated lateral control. In addition to the longitudinal control afforded by the ACC, the vehicle automatically follows the road curvature. In addition to that, lane changes may be performed in response to driver commands, for example, by flicking the indicator stalk in the direction of the lane change when in highly automated driving mode, much like a function of the most recent addition of automation on the market, the Tesla Motors (2016) Autopilot Lane change functionality (however, in its current state the automation does not assess the traffic in the adjacent lane before changing lane). The longitudinal and lateral automation subroutines are further explained in the sections that follow.

3.2.1   LONGITUDINAL AUTOMATION

The primary function and fundamental requirement of a longitudinal control system is to control and adapt speed to leading vehicles and/or driver settings (this is referred to as target velocity, v_{target}). Further requirements and assumptions for a longitudinal (i.e. ACC-based) control system are that

1. The system cannot be engaged while the vehicle is driving in reverse (Vakili, 2015).
2. The driver has the ability to override the system.
3. The system must ensure that the velocity set by the driver is maintained in the absence of a slow leading vehicle.
4. The system uses acceleration thresholds to ensure comfortable driving during normal operating conditions.
5. The system must ensure that the vehicle is slowed down to the velocity of a slower moving lead-vehicle and that the desired headway is maintained.

6. The system ignores deceleration thresholds when such a threshold hinders bringing the vehicle to a safe system state through slowing down or stopping completely.
7. The system hands back control to the driver when the operational limits are approached. These limits include sensor failure, geographical constraints or external factors leading to degraded system performance (this can be simulated through a shutdown event specified in an event file of the automation toolbox).

In the algorithm, the leading vehicle is considered to be a vehicle driving in the host vehicle's lane within the range of the simulated radar. The algorithm to access vehicle data for the lead-vehicle is described in Eriksson and Stanton (2017a). Parameters related to the lead-vehicle are denoted with the subscript *lead*. The longitudinal control algorithm is designed as a Finite State Machine (FSM), containing three states (cruise, follow and adapt), each with its own controller characteristics. There are several conditions that need to be fulfilled before the FSM can transition from one state to another, the process of determining controller states are shown in Pseudocode 1.

```
If Detected (lead Vehicle) Then
  If dlead <100 m ∧ dlead > 0 Then

    If Vlead < Vtarget ∧ Vhost > 0 Then
      If dlead/Vhost < thw_desired x 1.15 Then
       Vehicle detected - in range - Following
      Else If |Vtarget-Vhost| < 3.5 m/s Then
       Vehicle detected - in range - Cruising
      Else
       Vehicle detected - in range - Adapting speed
      End If
     Else
      If |Vtarget-Vhost| > 3.5 m/s Then
       Vehicle detected - too fast - Adapting speed
      Else
       Vehicle detected - too fast - Cruising
      End If
     End If
    Else
     If |Vtarget-Vhost| > 3.5 m/s Then
      Vehicle detected - out of range- Adapting speed
     Else
      Vehicle detected - out of range - Cruising
     End If
    End If
   Else
   If |Vtarget-Vhost| > 3.5 m/s Then
    No vehicle detected - Adapting speed
   Else
    No vehicle detected - Cruising
   End If
  End If
```

Pseudocode 1: the algorithm to determine the control mode of the longitudinal control subroutine. Note: ^ is the logical 'AND' operator.

Each state has a proportional–integral–derivative (PID) controller and, depending on the state, different gains for the different parameters are used. The transfer function of a PID controller is the sum of the outputs of three sub-controllers: a proportional, an integral and a derivative controller. The error signal undergoes processing in each controller, the resulting signals are added and they constitute the total output from the PID controller.

In the case of the automation, one of the inputs to the control system is the target velocity, and the output is a number representing the 'counts' of the virtual pedal position. Positive output values are signals sent to the virtual throttle pedal.

For negative signals, their absolute values represent the virtual brake pedal position. The feedback signal is the current vehicle's velocity (the outcome of the process).

Because the environment is inherently digital, discrete mathematics applies in the computations. Hence, the time differential resolution (Δt) is limited to a single simulation frame, that is, 1/30 for a frequency of 30 Hz. The controller's output (the number of counts) is governed by Equation 3.1, where $e_i = \Delta v_i = v_{target}, i - v_i$ is the error term representing the difference between the set velocity and vehicles current velocity at the current instant of time, i. Equation 3.1 represents the PID controller for car automation in a discrete simulation environment:

$$n_{pedal} = K_P e_i + K_I \Delta t \sum_{j=0}^{i} e_j + K_D \cdot \frac{e_i - e_{i-1}}{\Delta t} \qquad (3.1)$$

The K-coefficients of the sub-controllers represent their gains and have a significant impact on the behaviour of the system, as their relative and absolute values determine the lead time, overshoot or damping characteristics. Therefore, different control states require different controllers.

3.2.2 CONTROL STATES

3.2.2.1 Follow
The Follow state aims to maintain a constant time-headway to the lead-vehicle. This is a more challenging task than maintaining velocity, as the distance is controlled by the host vehicle velocity relative to the lead-vehicle. The time-headway is set by the driver and is defined as $t = d_{lead}/V_{host}$.

3.2.2.2 Adapt
The Adapt state is used for smooth velocity adjustments to meet the desired target velocity. The velocity error signal in the Adapt state is defined differently than in other states. The ultimate target velocity is still either the velocity set by the driver or the externally limited velocity (i.e. coming from a slower leading vehicle). However, in order to attain a smooth manoeuvre and velocity adjustment, the error signals refer to instantaneous target velocity, which comes from linear interpolation from the vehicle current velocity and the target velocity.

The interpolated velocity is calculated by the utilisation of Bezier curves. Bezier curves are frequently used in computer graphics to render animations or vector graphics. The Bezier curves are used to draw smooth curves that can be scaled dynamically and indefinitely. In animation Bezier curves can be used to control the velocity over time of animated objects. These characteristics make Bezier functions well suited for use in trajectory planning and interpolation. Bezier functions have been proposed as a way of planning and traversing trajectories in a two-dimensional space by Choi et al. (2009). Such an algorithm would be divided into two parts, trajectory planning and trajectory interpolation (Choi et al., 2009). In the current implementation of the control algorithm for longitudinal control, the Bezier functions are used to interpolate the velocity of the host vehicle to a set velocity or a leading vehicle's velocity to ensure smooth acceleration and deceleration by using a first-order Bezier (see Equation 3.2). The following is the equation for a first-order Bezier curve:

$$\text{First - order Bezier} : (1-t)P_0 + t \times P_1, t \in [0,1] \tag{3.2}$$

where:
P_0 is the host vehicle velocity at the start of interpolation
P_1 is the target velocity

To plan the velocity trajectory, a manoeuvre duration must be computed to match the host vehicle velocity with the target velocity taking a 'comfortable acceleration' threshold, as shown in Equation 3.3. Following the computation of the manoeuvre duration, the time interval needed is rescaled to a value range between 0 and 1 taking the simulator frame rate into account through Equation 3.4. Equation 3.3 is the formula used for finding manoeuvre duration used for interpolation and Equation 3.4 represents the Rescaling of T_{Manouvre} to a scale of 0-1 based on the frequency of the simulator.

$$T_{\text{manouvre}} = \frac{\Delta v_i}{a_{\text{comfortable}}} \tag{3.3}$$

$$t = \frac{T_{\text{curr}} - T_{\text{interp.,start}}}{\left(T_{\text{manouvre}} \times Hz_{\text{simulation}}\right)} \tag{3.4}$$

where:
T_{curr} refers to the current time
$T_{\text{interp., start}}$ refers to the start of the interpolation time
$T_{\text{manoeuvre}}$ refers to the manoeuvre time calculated in Equation 3.3

When the manoeuvre duration has been determined and scaled to the appropriate range, the current velocity, target velocity and time is introduced to Equation 3.2 to create the trajectory. Following the creation of the trajectory, the controller set point interpolates along the trajectory until the target velocity is reached. This ensures that the acceleration threshold is never exceeded. An example of an interpolation between two points is shown in Figure 3.3.

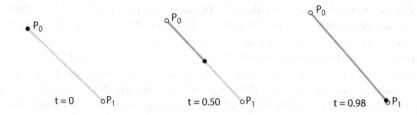

FIGURE 3.3 A first-order Bezier curve interpolating between P_0 and P_1 as specified in Equation 3.3. *Source*: Wikimedia Commons.

As the velocity is computed at each discrete step of the simulation, the error signals for the PID is significantly smaller, which is more manageable by the PID. The error signal for the PID is given by $e_i = \Delta v_i = v_i - v_{curr}$, which results in smoother acceleration and deceleration.

3.2.2.3 Cruise

The cruise state is used when the vehicle does not need to adjust its velocity more than 3.5 m/s (i.e. when small adjustments to the throttle output is required to maintain the set speed, when passing through hilly areas or curves) and when there is no lead-vehicle or a lead-vehicle faster than the set speed. The cruise state controls the velocity in accordance with Equation 3.1. The error term used for the PID controller is calculated as $e_i = v_{target} - v_{curr}$.

3.2.3 LATERAL AUTOMATION

The lateral control is responsible for steering the car and controlling its position in the desired lane. This is achieved by controlling the vehicle's lateral position with respect to the roads centreline and the centre of the desired lane. The target position is typically the exact coordinate of the centre of the lane. As steering is a non-linear problem, due to the effect of vehicle velocity on steering efficiency (the STISIM vehicle dynamics model has got some understeer at higher velocities), a stronger signal must be produced at higher velocities to follow a road's curvature. Therefore, the PID was modified to vary the proportional gain of the control signal as a function of current vehicle velocity. Equation 3.5 is the PID controller for the lateral controller.

$$n_{steering} = \left(K_{1P} + K_{2P} \cdot v_i^{K_{3P}} \right) \times e_i + K_I \Delta t \sum_{j=0}^{i} e_j + K_D \times \frac{e_i - e_{i-1}}{\Delta t} \qquad (3.5)$$

where:

K_{1p} is the main proportional scaling factor
K_{2p} is the second scaling factor for the effect of vehicle velocity on steering output
V_i is the current velocity of the host vehicle

e_i is the error term (the difference between current lane position and the lane centre)

K_I is the integral scaling factor

e_j is the integrated error term

K_D is the derivative scaling factor

3.3 ALGORITHM PERFORMANCE

The algorithms in this chapter were initially tested for comfort by human drivers in terms of following behaviour and steering performance to ensure that smooth manoeuvres were carried out in a fashion that ensured that participants would feel comfortable in the simulator. This was carried out in an unstructured manner throughout the development of the software toolkit, in a desktop environment in the early stages of development, followed by testing in the full simulator mock-up. The subjective testing showed that participants experienced the automation more aggressively in the full simulator mock-up than on the desktop simulation. This was attributed to screen size and perspective (in terms of viewing distance and object saliency) and was addressed before the software was deployed in the main mock-up to optimise car-following behaviour and controller characteristics to ensure a smooth experience.

In addition to subjective testing for comfort, many objective tests were carried out in order to demonstrate the effectiveness of the automated driving toolbox. These tests are detailed in the subsequent sections.

3.3.1 CAR-FOLLOWING

In order to assess longitudinal driving performance, the algorithms were run through a motorway driving scenario where a number of cars moved into the host vehicle's lane, as well as cut-ins as part of double lane changes.

Figure 3.4 shows the velocity profile of the host vehicle in relation to the set speed whereas Figure 3.5 shows the time-headway of any vehicles in front, in relation to the set time-headway (1.5 seconds).

As shown in Figure 3.5, the host vehicle closes the gap between the lead and host vehicle down to the desired time-headway and then maintains the desired time-headway consistently. When the lead-vehicle is no longer detected, the vehicle then returns to the original set speed. As Figure 3.4 shows, the host vehicle slows down below the lead-vehicle speed to accommodate the large need for sudden deceleration to achieve the desired time-headway when there is a large difference in velocity between the host and the lead-vehicle.

3.3.2 LANE-KEEPING

The same motorway scenario was used to assess the automated lateral control of the algorithm. Figure 3.6 shows the lateral deviation from the lane centre. It is possible to identify where the vehicle encountered a turn based on the deviation data; however,

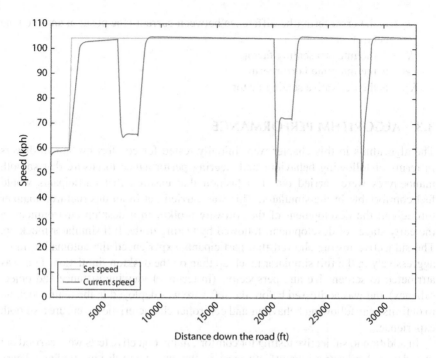

FIGURE 3.4 Velocity profile of motorway drive with car-following.

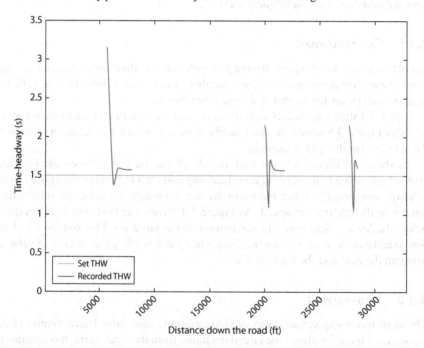

FIGURE 3.5 Time-headway profile of the car-following behaviour during motorway driving.

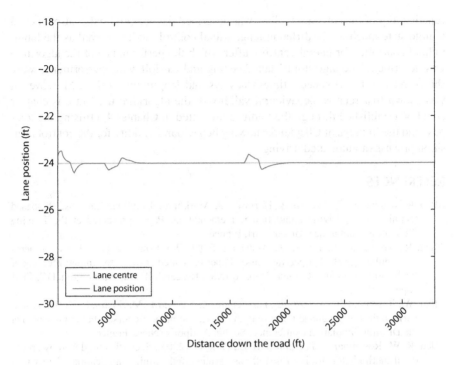

FIGURE 3.6   Lane-keeping performance (12-feet lane width) during a motorway drive.

the lateral deviation is at most ~15 centimetres from vehicle centre to lane centre, indicating good lateral vehicle control.

3.3.3 BEHAVIOURAL VALIDITY

The software toolbox presented in this chapter has been used in research conducted as part of Chapter 4 (in the same scenario as presented in Sections 3.3.1 and 3.3.2), which indicates relative behavioural validity when contrasted with on-road driving behaviour in a vehicle offering contemporary automated driving, as discussed in Chapter 5. Consequently, it lends validity to the use of the algorithms presented here for use in research into automated vehicles being conducted in simulators.

3.4 SUMMARY

This chapter introduced driving simulators as a safe way of testing driving tasks without putting drivers and other road users in harm's way. It then made the case for the use of driving simulators in automated driving research, whilst identifying a lack of easily implementable algorithms to enable such research in a dynamic way. This chapter sought to address this lack of software enabling research into the Human Factors of automated driving by disseminating a software toolbox used to enable automated driving in the Southampton University Driving Simulator. The general functionality of the software toolbox was described (i.e. longitudinal controllers and a finite state machine for different longitudinal control modes, as well as the longitudinal controller for lateral control), after which the performance of the algorithm was described. The algorithms' lane-keeping and car-following performance were shown relative to set speeds, time-headways and longitudinal offset. Moreover, it was shown that relative behavioural validity of the algorithm used in this chapter could be established through the findings presented in Chapter 4, utilising the toolbox, and the findings of Chapter 5, showing behavioural validity for the control transition process of automated driving.

REFERENCES

Allen, R., Rosenthal, T., Aponso, B., Harmsen, A. Markham, S. (1999a). Low cost, PC-based techniques for driving simulation implementation. Paper presented at the Driving Simulation Conference, Guyancourt, France.

Allen, R., Rosenthal, T., Hogue, J., Anderson, F. (1999b). Low-cost virtual environments for simulating vehicle operation tasks. Paper presented at the 78th Annual Meeting of the National Research Council Transportation Research Board, Washington, DC. TRB paper no 991136.

Allen, W., Rosenthal, T., Aponso, B., Parseghian, Z., Cook, M., Markham, S. (2001). A scenario definition language for developing driving simulator courses. Paper presented at the Driving Simulation Conference, Sophia Antipolis (Nice), France.

Allen, R. W., Rosenthal, T. J., Aponso, B. L., Park, G. (2003). Scenarios produced by procedural methods for driving research, assessment and training applications. Paper presented at the DSC North America 2003, Dearborn, MI.

Bainbridge, L. (1983). Ironies of automation. *Automatica*, vol 19, no 6, pp 775–779.

Boer, E. R., Penna, M. D., Utz, H. Pedersen, L., Sierhuis, M. (2015, 16–18 September). The role of driving simulators in evaluating autonomous vehicles. Paper presented at the Driving Simulation Conference, Max Planck Institute for Biological Cybernetics, Tübingen, Germany.

Carsten, O., Jamson, A. H. (2011). Driving simulators as research tools in traffic psychology. In B. Porter (Ed.), *Handbook of Traffic Psychology* (Vol 1). London: Academic Press, pp 87–96).

Casner, S. M., Hutchins, E. L., Norman, D. (2016). The challenges of partially automated driving. *Communications of the ACM*, vol 59, no 5, pp 70–77.

Choi, J. W., Curry, R., Elkaim, G. (2009). Path planning based on Bezier curve for autonomous ground vehicles. In *Proceedings of Wcecs 2008: Advances in Electrical and Electronics Engineering - Iaeng Special Edition of the World Congress on Engineering and Computer Science,* San Francisco, CA, pp 158–166.

De Winter, J., Happee, R., Martens, M. H., Stanton, N. A. (2014). Effects of adaptive cruise control and highly automated driving on workload and situation awareness: A review of the empirical evidence. *Transportation Research Part F: Traffic Psychology and Behaviour*, vol 27, pp 196–217.

De Winter, J. C. F., van Leeuwen, P., Happee, R. (2012). Advantages and disadvantages of driving simulators: A discussion. Paper presented at Measuring Behavior, Utrecht, the Netherlands, August 28–31.

De Winter, J., Wieringa, P., Dankelman, J., Mulder, M., Van Paassen, M., De Groot, S. (2007). Driving simulator fidelity and training effectiveness. Paper presented at the 26th European Annual Conference on Human Decision Making and Manual Control, Lyngby, Denmark.

Eriksson, A., Banks, V. A., Stanton, N. A. (2017). Transition to manual: Comparing simulator with on-road control transitions. *Accident Analysis & Prevention*, vol 102, pp 227–234.

Eriksson, A., Stanton, N. A. (2017a). he1y13/TrafficQuery_STISIM3: A traffic querying class for STISIM Drive Simulation Kernel – Build 3.07.04. Zenodo.

Eriksson, A., Stanton, N. A. (2017b). Takeover time in highly automated vehicles: Noncritical transitions to and from manual control. *Human Factors*, vol 59, no 4, pp 689–705.

Flach, J., Dekker, S., Stappers, P. J. (2008). Playing twenty questions with nature (the surprise version): Reflections on the dynamics of experience. *Theoretical Issues in Ergonomics Science*, vol 9, no 2, pp 125–154.

Funke, G. J. (2007). *The Effects of Automation and Workload on Driver Performance, Subjective Workload, and Mood*. University of Cincinnati.

Funke, G. J., Matthews, G., Warm, J. S., Emo, A., Fellner, A. N. (2016). The influence of driver stress, partial-vehicle automation, and subjective state on driver performance. In *Proceedings of the Human Factors and Ergonomics Society Annual Meeting*, vol 49, no 10, pp 936–940.

Godley, S. T., Triggs, T. J., and Fildes, B. N. (2002). Driving simulator validation for speed research. *Accident Analysis and Prevention*, vol 34, no 5, pp 589–600.

Greenshields, B. D. (1936). Reaction time in automobile driving. *Journal of Applied Psychology*, no 20, pp 353–358.

Hollnagel, E. and Woods, D. D. (1983). Cognitive systems engineering: New wine in new bottles. *International Journal of Man-Machine Studies*, vol 18, no 6, pp 583–600.

Jamson, A. H., Merat, N., Carsten, O. M. J., Lai, F. C. H. (2013). Behavioural changes in drivers experiencing highly-automated vehicle control in varying traffic conditions. *Transportation Research Part C-Emerging Technologies*, vol 30, pp 116–125.

Kircher, K., Eriksson, O., Forsman, Å., Vadeby, A., Ahlstrom, C. (2017). Design and analysis of semi-controlled studies. *Transportation Research Part F: Traffic Psychology and Behaviour*, vol 46, pp 404–412.

Kircher, K., Larsson, A., Hultgren, J. A. (2014). Tactical driving behavior with different levels of automation. *IEEE Transactions on Intelligent Transportation Systems*, vol 15, no 1, pp 158–167.

Kyriakidis, M., de Winter, J. C. F., Stanton, N., Bellet, T., van Arem, B., Brookhuis, K. et al. (2017). A human factors perspective on automated driving. *Theoretical Issues in Ergonomics Science*, pp 1–27.

Large, D. R., Burnett, G., Bolton, A. (2017). Augmenting landmarks during the head-up provision of in-vehicle navigation advice. *International Journal of Mobile Human Computer Interaction (IJMHCI)*, vol 9, no 2, pp 18–38.

Large, D. R., Crundall, E., Burnett, G., Harvey, C., Konstantopoulos, P. (2016). Driving without wings: The effect of different digital mirror locations on the visual behaviour, performance and opinions of drivers. *Applied Ergonomics*, vol 55, pp 138–148.

McIlroy, R. C., Stanton, N. A., Godwin, L., Wood, A. P. (2016). Encouraging eco-driving with visual, auditory, and vibrotactile stimuli. *IEEE Transactions on Human-Machine Systems*, vol 47, no 5, pp 661–672.

Miles, T. (2014, May 19). Cars could drive themselves sooner than expected after European push. Retrieved from https://www.reuters.com/article/us-daimler-autonomous-driving/cars-could-drive-themselves-sooner-than-expected-after-european-push-idUSKBN-0DZ0UV20140519.

Naujoks, F., Purucker, C., Neukum, A., Wolter, S., Steiger, R. (2015). Controllability of partially automated driving functions – Does it matter whether drivers are allowed to take their hands off the steering wheel? *Transportation Research Part F: Traffic Psychology and Behaviour*, vol 35, pp 185–198.

Neubauer, C. (2011). *The Effects of Different Types of Cell Phone Use, Automation and Personality on Driver Performance and Subjective State in Simulated Driving.* University of Cincinnati electronic thesis or dissertation. https://etd.ohiolink.edu/.

Neubauer, C., Matthews, G., Langheim, L., Saxby, D. (2012). Fatigue and voluntary utilization of automation in simulated driving. *Human Factors*, vol 54, no 5, pp 734–746.

Neubauer, C., Matthews, G., Saxby, D. (2014). Fatigue in the automated vehicle do games and conversation distract or energize the driver? In *Proceedings of the Human Factors and Ergonomics Society Annual Meeting*, vol 58, no 1, pp 2053–2057.

Neubauer, C., Matthews, G., Saxby, D. (2016). The effects of cell phone use and automation on driver performance and subjective state in simulated driving. In *Proceedings of the Human Factors and Ergonomics Society Annual Meeting*, vol 56, no 1, pp 1987–1991.

Nilsson, L. (1993). Behavioural research in an advanced driving simulator-experiences of the VTI system. In *Proceedings of the Human Factors and Ergonomics Society Annual Meeting*, vol 37, no 9, pp 612–616.

Nilsson, L. (1995). Safety effects of adaptive cruise control in critical traffic situations. Paper presented at the the Second World Congress on Intelligent Transport Systems: 'Steps Forward', Yokohama, Japan.

OpenDS. (2017). Retrieved from https://www.opends.eu/.

Park, G. D., Allen, R. W., Rosenthal, T. K. (2015). Novice driver simulation training potential for improving hazard perception and self-confidence while lowering speeding risk attitudes for young males. In *Proceedings of the Eighth International Driving Symposium on Human Factors in Driver Assessment, Training and Vehicle Design*, Salt Lake City, UT. Iowa City, IA: Public Policy Center, University of Iowa, pp 247–253.

SAE J3016. (2016). Taxonomy and definitions for terms related to driving automation systems for on-road motor vehicles, *J3016_201609*: SAE International.

Santos, J., Merat, N., Mouta, S., Brookhuis, K., de Waard, D. (2005). The interaction between driving and in-vehicle information systems: Comparison of results from laboratory, simulator and real-world studies. *Transportation Research Part F-Traffic Psychology and Behaviour*, vol 8, no 2, pp 135–146.

Sarter, N. B., Woods, D. D., Billings, C. E. (1997). Automation surprises. In G. Salvendy (Ed.), *Handbook of Human Factors and Ergonomics* (2nd edn.). New York: Wiley, pp 1926–1943.

Saxby, D. J., Matthews, G., Hitchcock, E. M., Warm, J. S. (2016). Development of active and passive fatigue manipulations using a driving simulator. In *Proceedings of the Human Factors and Ergonomics Society Annual Meeting*, vol 51, no 18, pp 1237–1241.

Saxby, D. J., Matthews, G., Hitchcock, E. M., Warm, J. S., Funke, G. J., Gantzer, T. (2008). Effect of active and passive fatigue on performance using a driving simulator. *Proceedings of the Human Factors and Ergonomics Society Annual Meeting*, vol 52, no 21, pp 1751–1755.

Seppelt, B. D. and Victor, T. W. (2016). Potential solutions to human factors challenges road vehicle automation. In G. Meyer and S. Beiker (Eds.), *Road Vehicle Automation 3*. Cham, Switzerland: Springer, pp 131–148.

Stanton, N. A. (2015, March). Responses to autonomous vehicles. *Ingenia*, vol 9, p 9.

Stanton, N. A., Marsden, P. (1996). From fly-by-wire to drive-by-wire: Safety implications of automation in vehicles. *Safety Science*, vol 24, no 1, pp 35–49.

Stanton, N. A., Young, M. S. (1998). Vehicle automation and driving performance. *Ergonomics*, vol 41, no 7, pp 1014–1028.

Stanton, N. A., Young, M. S., McCaulder, B. (1997). Drive-by-wire: The case of mental workload and the ability of the driver to reclaim control. *Safety Science*, vol 27, no 2–3, pp 149–159.

Stanton, N. A., Young, M. S., Walker, G. H., Turner, H., Randle, S. (2001). Automating the driver's control tasks. *International Journal of Cognitive Ergonomics,* vol 5, no 3, pp 221–236.

Tesla Motors. (2016). Model S Software Version 7.0.Retrieved from https://www.teslamotors.com/presskit/autopilot.

Underwood, G., Crundall, D., Chapman, P. (2011). Driving simulator validation with hazard perception. *Transportation Research Part F-Traffic Psychology and Behaviour*, vol 14, no 6, pp 435–446.

Vakili, S. (2015). Design and formal verification of an adaptive cruise control plus (acc+) system. Thesis submitted to the Department of Computing and Software and the School of Graduate Studes, McMaster University, Hamilton, ON.

van Winsum, W., Brookhuis, K. A., de Waard, D. (2000). A comparison of different ways to approximate time-to-line crossing (TLC) during car driving. *Accident Analysis & Prevention*, vol 32, no 1, pp 47–56.

Young, M. S., Stanton, N. A. (2007). Back to the future: Brake reaction times for manual and automated vehicles. *Ergonomics*, vol 50, no 1, pp 46–58.

4 Take-Over Time in Highly Automated Vehicles

When using a driver assistance system that is able to automate the driving task to such an extent that hands- and feet-free driving is possible (Society of Automotive Engineers/SAE Level 3, SAE J3016, 2016), the driver becomes decoupled from the operational and tactical levels of control (Michon, 1985; Stanton and Young, 2005), leaving the high-level strategic goals to be dealt with by the driver (until the point of resuming manual control). This is a form of 'driver-initiated automation', where the driver is in control of when the system becomes engaged or disengaged (Banks and Stanton, 2015, 2016; Lu and de Winter, 2015). Indeed, according to Bainbridge (1983), two of the most important tasks for humans in automated systems are monitoring the system to make sure it performs according to expectations and being ready to resume control when the automation deviates from expectation (Stanton and Marsden, 1996). Research has shown that vehicle automation has a negative effect on mental workload and situation awareness (Endsley and Kaber, 1999; Kaber and Endsley, 1997; Stanton and Young, 2005; Stanton et al., 1997; Young and Stanton, 2002) and that reaction times increase as the level of automation increases (Young and Stanton, 2007). This becomes problematic when the driver is expected to regain control when system limits are exceeded, as a result of a sudden automation failure. Failure-induced transfer of control has been extensively studied (see Desmond et al., 1998; Molloy and Parasuraman, 1996; Stanton et al., 1997, 2001; Strand et al., 2014; Young and Stanton, 2007). One failure-induced control-transition scenario Stanton et al. 1997) found that more than a third of drivers failed to regain control of the vehicle following an automation failure whilst using adaptive cruise control (ACC). Other research has shown that it takes approximately one second for a driver manually driving to respond to an unexpected and sudden braking event in traffic (Summala, 2000; Swaroop and Rajagopal, 2001; Wolterink et al., 2011). Young and Stanton (2007) report brake reaction times of 2.13 ± 0.55 seconds for drivers using ACC (SAE Level 1), and brake reaction times of 2.48 ± 0.66 seconds for drivers with ACC and assistive steering (SAE Level 2). By contrasting the results from Young and Stanton (2007) where drivers experienced an automation failure whilst a lead-vehicle suddenly braked, with Summala (2000) it seems like it takes an additional 1.1–1.5 s to react to sudden events requiring braking whilst driving with driver assistance automation (SAE Level 1) and partial driving automation (SAE Level 2). This increase, in combination with headways as short as 0.3 seconds (Willemsen et al., 2015) coupled with evidence that drivers are poor monitors (Molloy and Parasuraman, 1996), could actually cause accidents.

Evidently, automating the driving task seems to have a detrimental effect on driver reaction time (Young and Stanton, 2007). Therefore, as Cranor (2008), Chapter 2, and Eriksson and Stanton (2016) proposed, the driver needs to receive appropriate feedback if they are to successfully re-enter the driving control loop. According to Petermeijer et al. (2015), Zeeb et al. (2015) and Kerschbaum (2015), resuming control from an automated vehicle involves both physical and mental components. The driver resuming control must (1) shift visual attention from the current task to the road, (2) scan the driving scene to cognitively process and evaluate the traffic situation and select an appropriate action, (3) move hands and feet to the steering wheel and pedals so that control inputs can be made and (4) implement the appropriate action by actuating on the steering wheel and/or pedals as shown in Figure 4.1.

A survey of the literature on the control resumption process was carried out to gain an understanding of how long drivers need to resume manual control in an automated vehicle. Google Scholar and Web of Science were used to carry out the search using these search terms: TOR, Take Over + automation, control transitions + automation, take-over request. The inclusion criteria were that the manuscript had to report either the lead time given between a request to intervene and a 'system boundary', or the response-time of the driver to the request to intervene. A number of manuscripts were included in the review despite missing this after receiving clarifying information from the manuscript authors. A total of 25 papers reported either Take-Over Request lead time (TORlt: the lead time from a 'take-over request' to a critical event, such as a stranded vehicle) or Take-Over reaction time (TOrt: the time it takes the driver to take back control of the vehicle from the automated system when a take-over request has been issued, referred to as 'take-over time' in Figure 4.1), and they were therefore included in the review (see Table 4.1). It was found that the recent research efforts that have been made to determine the optimal TORlt and TOrt showed times varying from 0 to 30 seconds for TORlt and 1.14 to 15 seconds for TOrt, as shown in Table 4.1.

The review showed that the mean TORlt was 6.37 ± 5.36 seconds (Figure 4.2) with a mean reaction time of 2.96 ± 1.96 seconds. The most frequently used TORlts tended to be: 3 seconds with a mean TOrt of 1.14 ± 0.45 (studies 2, 13, 14, 18, 22), 4 seconds with a mean TOrt of 2.05 ± 0.13 (studies 4, 8, 22), 6 seconds with a mean TOrt of 2.69 ± 2.21 (studies 5, 8, 23) and 7 seconds with a mean TOrt of 3.04 ± 1.6 (studies 1, 6, 9, 17, 19, 25), as shown in Figure 4.3.

TOrts stay fairly consistent around 2–3.5 seconds in most control transitions, with a few outliers, as seen in Figure 4.3. Belderbos (2015), Merat et al. (2014), Naujoks et al. (2014) and Payre et al. (2016) show longer TOrt compared with the rest of

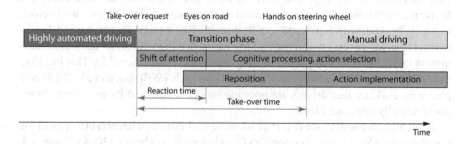

FIGURE 4.1   The take-over process as defined in Petermeijer et al. (2015).

TABLE 4.1

Papers Included in the Review

Paper		TORlt (s)	TOrt (s)	Modality
1	Gold et al. (2016)	7	2.47–3.61	—
2	Körber et al. (2015)	3	—	A
3	Louw et al. (2015b)	6.5	2.18–2.47	A
4	Zeeb et al. (2016)	2.5, 4	—	V
5	Damböck et al. (2012)	4, 6, 8	—	A
6	Kerschbaum et al. (2015)	7	2.22–3.09	V A
7	Belderbos (2015)	10	5.86–5.87	V A H
8	Walch et al. (2015)	4, 6	1.90–2.75	V A
9	Lorenz et al. (2014)	7	2.86–3.03	V A
10	Merat et al. (2014)	0	10–15	—
11	Naujoks et al. (2014)	—	2.29–6.90	V A
12	Schömig et al. (2015)	12	—	V A
13	Louw et al. (2015a)	3	—	V A
14	Zeeb et al. (2015)	2.5, 3, 3.5, 12	1.14	V A
15	Mok et al. (2015)	2, 5, 8	—	A
16	Gold et al. (2014)	5	1.67–2.22	V A
17	Radlmayr et al. (2014)	7	1.55–2.92	V A
18	Dogan et al. (2014)	3	—	V A
19	Gold et al. (2013)	5, 7	2.06–3.65	V A
20	van den Beukel and van der Voort (2013)	1.5, 2.2, 2.8	—	A
21	Melcher et al. (2015)	10	3.42–3.77	V A B
22	Naujoks and Nekum (2014)	0, 1, 2, 3, 4	—	V A
23	Feldhütter et al. (2016)	6	1.88–2.24	A
24	Payre et al. (2016)	2, 30	4.30–8.70	V A
25	Körber et al. (2016)	7	2.41–3.66	A

Modalities for the take-over request are coded as: A, Auditory; V, Visual; H, Haptic; and B, Brake Jerk.

the reviewed papers. Merat et al. (2014) and Naujoks et al. (2014) had the control transition initiated without any lead time, whereas Belderbos (2015) and Payre et al. (2016) provided drivers with a lead time. Merat et al. (2014) showed that there is a 10–15 second time lag between the disengagement of the automated driving system (ADS) and resumption of control by the driver. Notably, the control transition was system-initiated and lacked a pre-emptive take-over request which may have caused the increase in TOrt.

From a Gricean perspective (Chapter 2), the lack of information available to the driver to gain an understanding of why a sudden mode shift has occurred in Merat et al. (2014) is a violation of the maxims, and it could have contributed to the longer reaction time as more effort was needed to interpret the situation.

Similarly, Naujoks et al. (2014) observed a 6.9-second TOrt from when a take-over request was issued, and the automation disconnected until the driver resumed

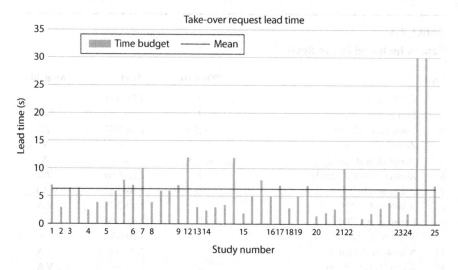

FIGURE 4.2 The TORlt used in the reviewed papers. Several papers used a multitude of take-over request lead times and thus contributed to several points of the graph.

control in situations where automation became unavailable due to missing line markings, the beginning of a work zone or entering a curve. Based on personal communication with the author, the vehicle would have crossed the lane markings after approximately 13 seconds and would have reached the faded lane markings approximately 10 seconds after the take-over request. The velocity in Naujoks et al. (2014) was 50 kph, which is fairly slow compared with most other take-over request studies that use speeds over 100 kph (studies 1, 3, 4, 6, 9, 10, 12, 14, 17, 19, 21, 23, 24, 25) and may have had an effect on the perceived urgency.

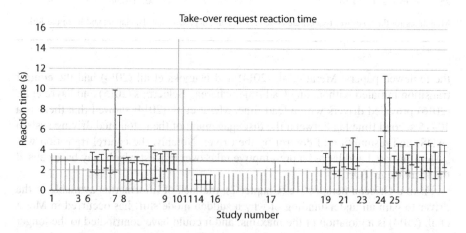

FIGURE 4.3 Take-Over reaction time averages for all the conditions in the reviewed studies. Some studies had more than one take-over event and are therefore featured multiple times.

Belderbos (2015) showed TOrts of 5.86 ± 1.57 to 5.87 ± 4.01 when drivers were given a TORlt of 10 seconds during unsupervised automated driving. Payre et al. (2016) utilised two different TORlts, 2 and 30 seconds.

These TORlts produced significant differences in TOrt, the 2-second TORlt produced a TOrt of 4.3 ± 1.2 seconds and the two scenarios that used the 30-second TORlt produced TOrts of 8.7 ± 2.7 seconds and 6.8 ± 2.5 seconds respectively. The shorter TOrt of the two 30-second take-over request events occurred after the 2-second emergency take-over request and could have been affected by the urgency caused by the short lead time in the preceding, shorter take-over request.

Merat et al. (2014) concluded, based on their observed TOrt, that there is a need for a timely and appropriate notification of an imminent control transition. This observation is in line with the current SAE guidelines, which state that the driver *'is receptive to a request to intervene and responds by performing dynamic driving task fallback in a timely manner'* (SAE J3016,2016, p. 20). In initial efforts to determine how long in advance the driver needs to be notified before a control transition is initiated, Damböck et al. (2012) and Gold et al. (2013) explored a set of take-over request lead times. Damböck et al. (2012) utilised three TORlts, 4, 6 and 8 seconds, and found that given an 8-second lead time, drivers did not differ significantly from manual driving. This was confirmed by Gold et al. (2013), who reported that drivers need to be warned at least 7 seconds in advance of a control transition to safely resume control. These findings seem to have been the inspiration for the TORlt of some recent work utilising timings around 7 seconds (studies 1, 6, 9, 17).

A caveat for a number of the reviewed studies is that the lead time given in certain scenarios, such as disappearing lane markings, construction zones and merging motorway lanes, is surprisingly short, from 0 to 12 seconds (c.f. Table 5), and it will likely be longer in on-road use-cases (studies 4, 5, 11, 14, 15, 21). The reason for this is the increasing accuracy of contemporary GPS hardware and associated services such as Google Maps. Such services are already able to direct lane positioning whilst driving manually, as well as notifying drivers of construction zones and alternate, faster routes. Thus, there is no evident gain to having short lead times in such situations.

Several of the studies reviewed have explored the effect of take-over requests in different critical settings by issuing the take-over request immediately preceding a time critical event (studies 1, 2, 3, 4, 6, 7, 8, 9, 13, 16, 17, 19, 20, 23, 24, 25). These studies have explored how drivers manage critical situations in terms of driving behaviour, workload and scanning behaviour.

Whilst it is of utmost importance to know how quickly a driver can respond to a take-over request and what the shortest take-over request times are in emergencies, there is a paucity of research exploring the time it takes a driver to resume control in normal, non-critical, situations. We argue that if the design of normal, non-critical, control transitions are designed based on data obtained in studies utilising critical situations, there is a risk of unwanted consequences, such as drivers not responding optimally due to too short lead time (suboptimal responses are acceptable in emergencies as drivers are tasked with avoiding danger), drivers being unable to fully regain situation awareness, and sudden, dramatic, increases in workload.

Arguably, these consequences should not be present in every transition of control as it poses a safety risk for the driver as well as other road users. Therefore, the aim of this study is to establish driver take-over time in normal traffic situations when, for example, the vehicle is leaving its operational design domain (ODD), as these will account for most of the situations (Nilsson, 2014; SAE J3016, 2016). We also explore how take-over request take-over time is affected by a non-driving secondary task, as this was expected to increase the reaction time (Merat et al., 2012).

Moreover, none of the papers included in the review mentioned the time it takes drivers to transition from manual to automated driving. Gaining an understanding of the time required to toggle an ADS on is important in situations, such as entering an area dedicated to automated vehicles or engaging the automated driving mode in preparation for joining a platoon, as proposed by the Safe Road Trains for the Environment (SARTRE) project (Robinson et al., 2010). Therefore, the aim of this study was to establish the time it takes a driver to switch to automated driving when automated driving features become available. Ultimately, this research aims to provide guidance about the lead time required to get the driver back into, and out of, the manual vehicle control loop.

4.1 METHOD

4.1.1 PARTICIPANTS

Twenty-six participants (10 female, 16 male) between 20 and 52 years of age ($M = 30.27$ $SD = 8.52$) with a minimum of 1 year and an average of 10.57 years ($SD = 8.61$) of driving experience were asked to take part in the trial. Upon recruitment, participants were screened for experience with Advanced Driver Assistance Systems (ADAS) and had to have a valid license for driving in the United Kingdom. Upon recruiting participants, their informed consent was obtained. The study complied with the American Psychological Association Code of Ethics and had been approved by the University of Southampton Ethics Research and Governance Office (ERGO number 17771).

4.1.2 EQUIPMENT

The experiment was carried out in a fixed-based driving simulator located at the University of Southampton. The simulator was a Jaguar XJ 350 with pedal and steering sensors provided by Systems Technology Inc. as part of STISIM Drive® M500W Version 3 (http://www.stisimdrive.com/m500w) providing a projected 140° field of view. Rear-view and side mirrors were provided through additional projectors and video cameras (Figure 4.4).

The original Jaguar XJ instrument cluster was replaced with a 10.6" Sharp LQ106K1LA01B Laptop LCD panel connected to the computer via a RTMC1B LCD controller board to display computer-generated graphics components for take-over requests.

Bespoke software had to be created to replace the original instrument cluster with the digital instrument cluster solution. C# and the Windows Presentation Foundation framework (WPF) were used as the underlying architecture. An open source gauge

FIGURE 4.4 The Southampton University Driving Simulator from the driver's point of view. (Image credit: Daniël Heikoop.)

element (WPF Gauge 2013.5.27.1000, Phillips, 2013) for the WPF framework was acquired from Codeplex and integrated into a WPF application. Upon connecting the digital instrument cluster to STISIM, severe latencies were discovered when rendering the changing values of the dials. This could be traced back to the frequency of the calls to the rendering engine, and the fact that the entirety of each gauge was rendered whenever new data was sent. To mitigate these issues, the gauges were modified by importing them into Photoshop to extract the dial and the gauge separately. These elements were then imported into Microsoft Expression Blend to create a custom WPF controller. The controller accepted a value as an input and rotated the dial with a gain factor corresponding to the correct value on the gauge.

This reduced the amount of rendering needed, but it caused a jagged transformation when the dial rotated; this was mitigated by applying an animation to the rotation of the dial. The end result was a smoothly updating gauge that was incorporated in the full cluster set-up. The default configuration of the instrument cluster is shown in Figure 4.5.

When a take-over request was issued, the engine speed dial was hidden and the request was shown in its place. The symbol asking for control resumption is shown in Figure 4.6, and the symbol used to prompt the driver to re-engage the automation is shown in Figure 4.7.

FIGURE 4.5 The instrument cluster in its default configuration.

Please resume control

FIGURE 4.6   The take-over request icon shown on the instrument cluster. The icon was coupled with a computer-generated voice message stating, 'please resume control'.

Automation available

FIGURE 4.7   The icon is shown when the automation becomes available. The icon was coupled with a computer-generated voice message stating, 'automation available'.

The mode-switching human–machine interface was located on a Windows tablet in the centre console, consisting of two buttons, used either to engage or to disengage the automation. To enable dynamic disengagement and re-engagement of the automation, bespoke algorithms were developed and are reported in Chapter 3.

4.1.3   EXPERIMENT DESIGN

The experiment had repeated-measures, within-subject design with three conditions: Manual, Highly Automated Driving and Highly Automated Driving with a secondary task. The conditions were counterbalanced to counteract order effects. For the automated conditions, participants drove at 112.65 kph on a 30 kilometre, three-lane highway with some curves, with oncoming traffic in the opposing three lanes separated by a barrier and in moderate traffic conditions. The route was mirrored between the two automated conditions to reduce familiarity effects whilst keeping the roadway layout consistent. In the manual driving conditions, drivers drove a shortened version (10 minutes) of the route in Figure 4.8 under identical traffic and road layout conditions as those in the automated conditions. In the secondary task condition, drivers were asked to read (in their head) an issue of *National Geographic* whilst the ADS was engaged in order to remove them from the driving (and monitoring) task.

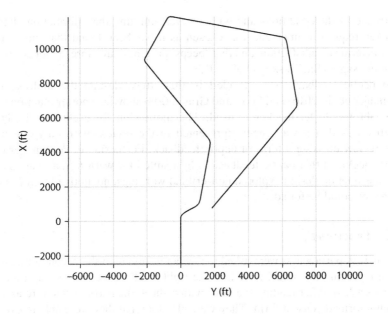

FIGURE 4.8 A bird's eye view of the road layout in the passive monitoring condition (route was mirrored for the secondary task condition).

During both conditions, drivers were prompted to either resume control from, or relinquish control to, the ADS. The drivers started in manual driving mode at the start of each automated driving condition and the timer to trigger the prompts to transition between control modes was triggered after 2 minutes of manual driving. Consequently, drivers spent between 2 minutes and 30 seconds and 2 minutes and 45 seconds in manual mode before being asked to engage the ADS (Figure 4.9). The interval in which the control transition requests were issued ranged from 30 to 45 seconds, thus allowing for approximately 24 control transitions, of which half were to manual control.

The control transition requests were presented as both a visual cue (c.f. Figures 4.6 and 4.7) and an auditory message, in line with previous research (studies 6, 7, 8, 9, 11, 12, 13, 14, 16, 17, 18, 19, 22, 24), in the form of a computer-generated, female voice stating, 'please resume control' or 'automation available'. From a Gricean perspective, the warnings were coherent with the maxims, as they conveyed sufficient information to indicate the need for a transition of vehicle control, in a way that ensured that the driver would notice them even when preoccupied with a reading

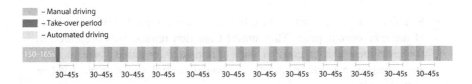

FIGURE 4.9 The take-over request interval utilised in Chapters 4, 5 and 6.

task in one of the conditions and without oversaturating the information channel. A spoken request to intervene was chosen as it has been found that this is rated higher in terms of satisfaction, and that 'beeps' or 'tones' are more apt for conveying urgent messages (Bazilinskyy et al., 2017).

No haptic feedback was included in this study, despite the findings from Petermeijer et al. (2016) and Scott and Gray (2008) showing shorter reaction times when vibrotactile feedback was used. The motivation for excluding the haptic modality was that it was under-represented in the review, with only one paper in the review utilising a form of haptic feedback. Furthermore, Petermeijer et al. (2016) concluded that haptic feedback is best suited for warnings, and as the current experimental design explored non-critical warnings, no motivation for including haptics could be found.

4.1.4 Procedure

Upon arrival, participants were asked to read an information sheet, containing information regarding the study, including the right to abort the trial at any point, no questions asked. After reading the information sheet, the participants were asked to sign an informed consent form. They were also told that they were able to override any system inputs via the steering wheel, throttle or brake pedals.

Drivers were reminded that they were responsible for the safe operation of the vehicle regardless of its mode (manual or automated) and thus needed to be able to safely resume control in case of failure. This is in accordance with current legislation (United Nations, 1968) and recent amendments to the Vienna Convention on Road Traffic. Participants were informed that the system may prompt them to either resume or relinquish control of the vehicle and that when such a prompt was issued, they were required to adhere to the instruction, but only when they felt safe doing so. This instruction was intended to reduce the pressure on drivers to respond immediately and to reinforce the idea that they were ultimately responsible for safe vehicle operation. Before the experimental drives, the participants drove a 5-minute familiarisation drive during which they experienced the automation, and how the human–machine interface (HMI) requested driver intervention. At the end of each driving condition, participants were asked to fill out the NASA-RTLX questionnaire (Byers et al., 1989). They were also offered a short break before continuing the study. Reaction time data were logged for each transition to and from manual control.

4.1.5 Dependent Variables

The following metrics for participants were collected for each condition:

- Reaction time to the control transition request was recorded from the onset of the take-over request. The control transition request was presented in the instrument cluster coupled with a computer-generated voice to initiate a change in mode to and from manual control and was recorded in milliseconds.

- The NASA raw TLX was used to evaluate the perceived workload (Byers et al., 1989; Hart and Staveland, 1988). The questionnaire covers six items: mental demand, physical demand, temporal demand, performance, effort and frustration. The items have a 21-tick Likert scale, ranging from 'very low' to 'very high', except the performance scale, which ranges from 'perfect' to 'failure'. The overall workload score was calculated by the summation of the individual item scores divided by six (Byers et al., 1989; Hart and Staveland, 1988).

4.2 ANALYSIS

The dependent measures were tested for normal distribution using the Kolmogorov–Smirnov test, which revealed that the data was non-normally distributed. Furthermore, as the TOrt data is reaction time data, the median TOrt values for each participant were calculated, after which the Wilcoxon signed-rank test was used to analyse the time and workload data. The box plots in Figures 4.10 and 4.12 had their outlier thresholds adjusted to accommodate the log-normal distribution of the TOrt data by using the LIBRA library for MATLAB® (Verboven and Hubert, 2005) and its method for robust box plots for non-normally distributed data (Hubert and Vandervieren, 2008). Effect sizes were calculated as: $r = \left| Z/\sqrt{N} \right|$. As this dataset is used in more than one chapter in this book, the significance (alpha) level in this chapter is set to 0.01.

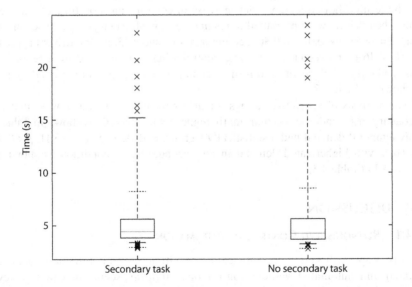

FIGURE 4.10 Adjusted box plot of control transition times from manual driving to Automated Driving. The dashed horizontal line indicates the max/min values assuming a normal distribution.

TABLE 4.2

Descriptive Statistics of the Control Transition Times from Automated Driving to Manual Control, and from Manual Control to Automated Driving

	Meta-Review	From Automated to Manual		From Manual to Automated	
		No Secondary Task	Secondary Task	No Secondary Task	Secondary Task
Median	2,470	4,567	6,061	4,200	4,408
IQR	1,415	1,632	2,393	1,964	1,800
Min	1,140	1,975	3,179	2,822	2,926
Max	15,000	25,750	20,994	23,884	23,221

Note: Times are in milliseconds. Descriptive statistics from the presented TOrts from the reviewed articles are also shown.

4.3 RESULTS

The results showed that it took approximately 4.2–4.4 ± 1.96–1.80 seconds (median) to switch to automated driving, see Table 4.2. No significant differences between the two conditions could be found when drivers transitioned from manual to automated driving ($Z = -0.673$, $p = 0.5$, $r = 0.13$). Control transition times from manual to automated driving in the two conditions is shown in Figures 4.10 and 4.11. Upon checking for learning effects after the repeated transitions of control, no learning effects could be found.

The results showed a significant increase in control transition time of ~1.5 seconds when drivers were prompted to resume control whilst engaged in a secondary task ($Z = -4.43$, $p < 0.01$, $r = 0.86$) see Figures 4.12 and 4.13. It took drivers approximately 4.46 ± 1.63 seconds to resume control when not occupied by a secondary task, and 6.06 ± 2.39 seconds to resume control when engaged in a secondary task as shown in Table 4.2.

The analysis of subjective ratings for driver mental workload showed that the secondary task condition has marginally higher scores overall, as shown in Table 7. Only temporal demand had a statistically significant difference ($Z = -3.11$, $p < 0.01$, $r = 0.61$), with higher rated demand in the secondary task condition as shown in Figure 4.14 (Table 4.3).

4.4 DISCUSSION

4.4.1 RELINQUISHING CONTROL TO AUTOMATION

In this chapter, drivers were subjected to multiple control transitions between manual and automated vehicle control in a highway scenario. Upon reviewing the literature, no mention of how long the driver takes to engage an ADS was found, making this study a first of its kind. It was found that drivers take between 2.82 and 23.8 seconds (Median = 4.2–4.4) to engage automated driving when the system indicated that the feature is available. No significant differences between the two conditions were found, but as Figure 4.10 shows, there was a large range in the time it takes to relinquish control. It is clear from Figure 4.11 that designing for the median or average driver effectively excludes a large part of the user group, which could have severe implications for drivers who fall outside of the mean or median. It has been common practice in Human Factors and Anthropometrics to design for 90% of the population, normally by accommodating the range between the 5th percentile female and the 95th percentile male (Porter et al., 2004).

Thus, it is important that vehicle manufacturers are made aware of the intra-individual differences, as such differences have a large effect on the larger traffic system

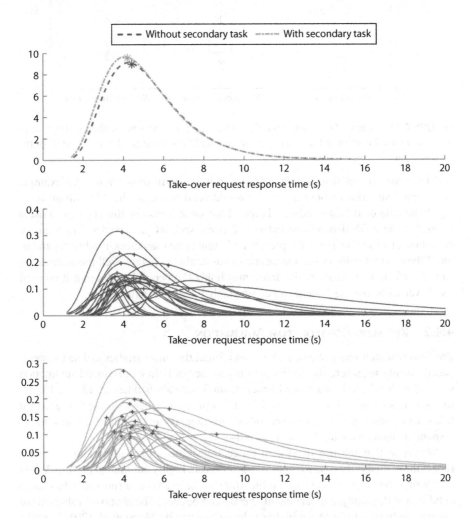

FIGURE 4.11   Top: A distribution plot of TOrt when drivers were prompted to engage the automation. The asterisk * marks the median value, the X axis contains 160 bins. Middle and bottom: Distribution plot of individual participants in the passive monitoring condition and secondary task condition respectively.

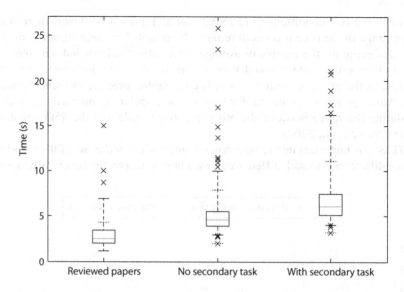

FIGURE 4.12 Adjusted box plot of the Take-Over reaction time when switching from automated to manual control in the two experimental conditions contrasted with the TOrt of the reviewed papers.

if drivers are expected to toggle ADSs within a certain time frame. An example of potential situations where the driver would need to toggle the ADS might be in highly automated driving-dedicated areas. Moreover, it may be that the time it takes to engage the ADS depends on external factors, such as perceived safety, weather conditions, traffic flow rates, the presence of vulnerable road users, roadworks and so on. If the driver deems a situation unsafe or has doubts as to how well the automation would perform in a situation, the driver may hold off on completing a transition until the driver feels that the system can comfortably handle the situation.

4.4.2 RESUMING CONTROL FROM AUTOMATION

Previous research was reviewed, and it was found that most studies utilised system-paced transitions, where the ADS warns in advance of failure or reduced automation support with relatively short lead times, from 3 seconds (studies 2, 13, 14, 18, 22) to 7 seconds (studies 1, 6, 9, 17, 19, 25). It has previously been shown that whilst it takes approximately 2.47 ± 1.42 seconds on average, it can take up to 15 seconds to respond to such an event (Merat et al., 2014).

We argue that the use of system-paced transitions, albeit important, does not reflect the primary use case for control transitions in highly automated driving. When comparing the range of TOrt in the literature to the user-paced (no secondary task) condition in this study, a great deal of overlap can be seen. The observed values in the current study are closer to the higher values observed by Merat et al. (2014), whilst the median range of 4.56–6.06 seconds is closer to the range of times suggested by Gold et al. (2013) and Damböck et al. (2012). It is evident that there is a large spread in the TOrt, which should be considered when designing driving automation, as the

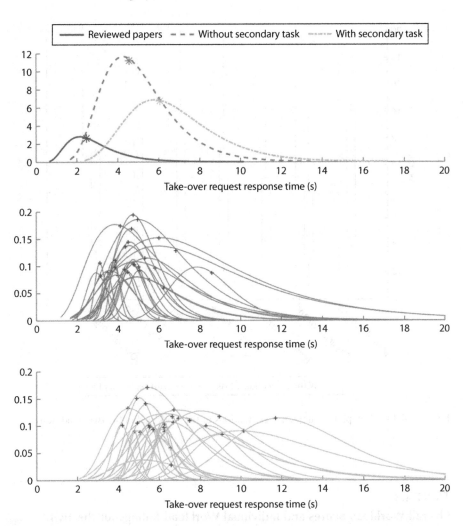

FIGURE 4.13 Top: A distribution plot of TOrt when drivers were prompted to resume manual control. The asterisk* marks the median value, the X axis contains 160 bins. Middle and Bottom: Distribution plot of individual participants in the passive monitoring condition and secondary task condition respectively.

range of performance is more important than the median or mean, as these exclude a large portion of drivers.

When subjecting drivers to take-over requests without time restrictions, it was found that drivers take between 1.97 and 25.75 seconds (Median=4.56) to resume control from automated driving in normal conditions, and between 3.17 and 20.99 seconds (Median=6.06) to do so whilst engaged in a secondary task preceding the control transition (Figure 4.13). This shows that there is a median 2-second difference in control transition times in the reviewed manuscripts compared with the user-paced control transitions. There was a large effect of secondary task engagement on TOrt, showing an increase in driver control resumption times when engaged in a secondary task.

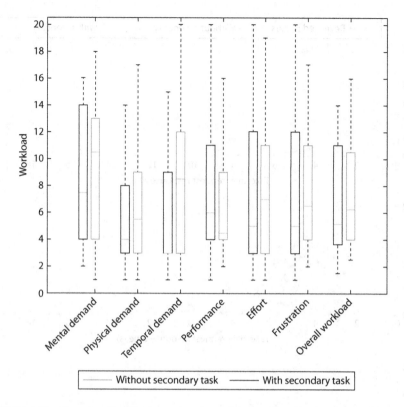

FIGURE 4.14   Box plot of subjective estimations of the workload in the two conditions.

TABLE 4.3
Overall Workload Scores and Individual Workload Ratings for the Two Conditions

Variable	Without Secondary Task Median (IQR)	With Secondary Task Median (IQR)	Z	p	r
Overall workload	5 (7.33)	6.2 (6.5)	−1.953	0.051	0.38
Mental demand	7.5 (10)	10.5 (9)	−1.41	0.16	0.28
Physical demand	4 (5)	5.5 (6)	−1.93	0.054	0.38
Temporal demand	3 (6)	8.5 (8)	−3.11	0.00**	0.61
Performance	6 (7)	4.5 (5)	−0.47	0.63	0.09
Effort	5 (9)	7 (8)	−0.23	0.82	0.05
Frustration	4 (9)	6.5 (7)	−1.04	0.3	0.2

** = Significant at the 0.01 level.

This might be explained in part by the nature of the secondary task, as the driver had to allocate time to put down the magazine they were asked to read whilst the automated driving feature was activated. It could also be partly attributed to driver task adaptation, by holding off transferring control until they have had time to switch between the reading task and driving task. This is supported by research indicating that drivers tend to adapt to external factors such as traffic complexity to allow for more time to make decisions (Eriksson et al., 2014), for example, by slowing down when engaged in secondary tasks (Cooper et al., 2009) or the expectation of resuming control (Young and Stanton, 2007). Eriksson et al. (2014) also found that drivers' information needs changed when TORlt and traffic situations were simulated in an online environment. In light of these results, there is a case for 'adaptive automation' that modulates TORlt by, for example, detecting whether the driver gaze is off the road for a certain time-period, providing the driver with a few additional seconds before resuming control. Further adaptations by the ADS could be to tailor the information presented to the driver to the available TORlt and traffic situation in accordance to the maxims presented in Chapter 2, as providing appropriate information (i.e. contextually relevant) has been shown to decrease TOrt (Beller et al., 2013).

Furthermore, the 1.5-second increase of control resumption time found when the driver was engaged in a reading task is similar to the reaction time increase caused by the introduction of automated driving features observed by Young and Stanton (2007) compared with reaction time in manual driving (Summala, 2000). Therefore, a further increase of reaction times when drivers are engaged in other tasks has to be expected, but measures must be taken to reduce the increase in reaction time, for example through the addition of informative displays, to reduce the risk of accidents as suggested in Chapter 2 (Cranor, 2008; Eriksson and Stanton, 2016).

There was a significant increase in perceived temporal demand when drivers were tasked with reading whilst the automation was engaged. This increase in perceived temporal demand may have been caused by the take-over request, and the driver not being fully sure of how long the vehicle could manage before a forced transition would occur (this was not a possibility in the current experiment).

This increase in perceived temporal demand could also be attributed to the pace of the experiment, and that the drivers were required to pick up and put down the magazine whenever a control transition was issued. Overall there are little differences in workload, and the median workload in both conditions was approximately at the halfway point on the scale, implying optimal loading (Stanton et al., 2011). These results indicate that the drivers were able to self-regulate the control transition process by adapting the time needed to resume control (Eriksson et al., 2014; Kircher et al., 2014) and therefore maintaining optimal levels of workload.

4.5 CONCLUSIONS

4.5.1 RELINQUISHING CONTROL TO AUTOMATION

The literature on control transitions in highly automated driving is absent in research reports on transitions from manual to automated vehicle control. In a first-of-a-kind study, it was found that it takes drivers between 2.8 and 23.8 seconds to switch from manual to automated control. This finding has some implications for the safety of drivers merging into automated driving-dedicated lanes, or other infrastructure whilst in manual mode. Such an event may require certain adaptations for traffic already occupying such a lane. Adaptations may include increasing time-headway or reducing speed to accommodate the natural variance in human behaviour to avoid collisions or discomfort for road users in such a lane. Moreover, it may be that part of the variance could be reduced by designing merging zones on straight, uncomplicated road sections as drivers may otherwise hold off transferring control to the ADS until the driver feels it safe to hand control to the ADS.

4.5.2   Resuming Control from Automation

A review of the literature found that most papers tend to report the mean TOrt and often fail to report standard deviation and range (c.f. Figure 4.3); thus, the variance in control transition times remains unknown. Additionally, the reviewed papers tend to give drivers a lead time of between 0 and 30 seconds between the presentation of a take-over request and a critical event, with the main part of the reviewed papers using a 3 or 7 second lead time. In this chapter, it was found that the range of time in which drivers resume control from the ADS was between 1.9 and 25.7 seconds, depending on task engagement.

The spread of TOrt in the two conditions in this study indicates that mean or median values do not tell the entire story when it comes to control transitions. Notably, the distribution of TOrt approaches platykurtic (c.f. Figure 4.13) when drivers are engaged in a secondary task. This implies that vehicle manufacturers must adapt to the circumstances, providing more time to drivers who are engaged in secondary tasks, whilst in the highly automated driving mode to avoid excluding drivers at the tail of the distribution. In light of this, designers of automated vehicles should not focus on the mean, or median, driver when it comes to control transition times.

Rather, they should strive to include the larger range of control transitions times, so they do not exclude users that fall outside the mean or median. Moreover, policymakers should strive to accommodate these inter- and intra-individual differences in their guidelines for 'sufficiently comfortable transition time' (National Highway Traffic Safety Administration, 2013). When drivers were allowed to self-regulate the control transition process, little differences could be found in workload between the two conditions. This lends further support to the argument for designing for the range of transition times rather than the mean or median in non-critical situations.

Lastly, based on the large decrease in TOrt kurtosis when drivers were engaged in a secondary task, it may also be the case that future automated vehicles need to adapt the TORlt to account for drivers engaged in other, non-driving tasks and even adapt TORlt to accommodate external factors, such as traffic density and weather.

4.6   FUTURE DIRECTIONS

Chapter 5 builds on the work of this chapter by continuing to investigate self-paced transitions of control. Chapter 5 focusses on validating the findings generated in the simulator in this chapter to ensure that the disseminated research is applicable to real-world driving scenarios. Moreover, it is hoped that the findings in Chapter 5 lend validity to the algorithms presented in Chapter 3 and used in this chapter. It is also hoped that these findings validate contemporary research into control transitions previously disseminated in the scientific literature.

REFERENCES

Bainbridge, L. (1983). Ironies of automation. *Automatica*, vol 19, no 6, pp 775–779.

Banks, V. A., Stanton, N. A. (2015). Contrasting models of driver behaviour in emergencies using retrospective verbalisations and network analysis. *Ergonomics*, vol 58, no 8, pp 1337–1346.

Banks, V. A., Stanton, N. A. (2016). Keep the driver in control: Automating automobiles of the future. *Applied Ergonomics*, vol 53, Pt B, pp 389–395.

Bazilinskyy, P., Eriksson, A., Petermeijer, B., de Winter, J. (2017). Usefulness and satisfaction of take-over requests for highly automated driving. Paper presented at the Road Safety and Simulation Conference, The Hague, the Netherlands.

Belderbos, C. (2015). Authority transition interface: a human machine interface for taking over control from a highly automated truck. Master's thesis. TU Delft, Delft University of Technology, Delft, the Netherlands.

Beller, J., Heesen, M., Vollrath, M. (2013). Improving the driver-automation interaction: An approach using automation uncertainty. *Human Factors*, vol 55, no 6, pp 1130–1141.

Byers, J. C., Bittner, A., Hill, S. (1989). Traditional and raw task load index (TLX) correlations: Are paired comparisons necessary. In A. Mital (Ed.), *Advances in Industrial Ergonomics and Safety I*. London: Taylor & Francis, pp 481–485.

Cooper, J. M., Vladisavljevic, I., Medeiros-Ward, N., Martin, P. T., Strayer, D. L. (2009). An investigation of driver distraction near the tipping point of traffic flow stability. *Human Factors*, vol 51, no 2, pp 261–268.

Cranor, L. F. (2008). A framework for reasoning about the human in the loop. In *Proceedings of the 1st Conference on Usability, Psychology, and Security*.

Damböck, D., Bengler, K., Farid, M., Tönert, L. (2012). Übernahmezeiten beim hochautomatisierten Fahren. *Tagung Fahrerassistenz. München*, vol 15, p 16.

Desmond, P. A., Hancock, P. A., Monette, J. L. (1998). Fatigue and automation-induced impairments in simulated driving performance. *Human Performance, User Information, and Highway Design*, vol 1628, pp 8–14.

Dogan, E., Deborne, R., Delhomme, P., Kemeny, A., Jonville, P. (2014). Evaluating the shift of control between driver and vehicle at high automation at low speed: The role of anticipation. Paper presented at the Transport Research Arena (TRA) 5th Conference: Transport Solutions from Research to Deployment.

Endsley, M. R., Kaber, D. B. (1999). Level of automation effects on performance, situation awareness and workload in a dynamic control task. *Ergonomics*, vol 42, no 3, pp 462–492.

Eriksson, A., Lindström, A., Seward, A., Seward, A., Kircher, K. (2014). Can user-paced, menu-free spoken language interfaces improve dual task handling while driving? In M. Kurosu (Ed.), *Human-Computer Interaction. Advanced Interaction Modalities and Techniques* (vol 8511). Cham, Switzerland: Springer, pp 394–405.

Eriksson, A., Stanton, N. A. (2016). The chatty co-driver: A linguistics approach to human-automation-interaction. Paper presented at the IEHF2016.

Feldhütter, A., Gold, C., Schneider, S., Bengler, K. (2016). How the duration of automated driving influences take-over performance and gaze behavior. Paper presented at the Arbeit in komplexen Systemen – Digital, vernetz, human?! 62. Kongress der Gesellschaft für Arbeitswissenschaft, Aachen, Germany.

Gold, C., Damböck, D., Lorenz, L., Bengler, K. (2013). "Take over!" How long does it take to get the driver back into the loop? In *Proceedings of the Human Factors and Ergonomics Society Annual Meeting*, vol 57, no 1, pp 1938–1942.

Gold, C., Korber, M., Lechner, D., Bengler, K. (2016). Taking over control from highly automated vehicles in complex traffic situations: The role of traffic density. *Human Factors*, vol 58, no 4, pp 642–652.

Gold, C., Lorenz, L., Bengler, K. (2014). Influence of automated brake application on take-over situations in highly automated driving scenarios. In *Proceedings of the FISITA 2014 World Automotive Congress* (in print).

Hart, S. G., Staveland, L. E. (1988). Development of NASA-TLX (Task Load Index): Results of empirical and theoretical research. *Advances in Psychology*, vol 52, pp 139–183.

Hubert, M., Vandervieren, E. (2008). An adjusted boxplot for skewed distributions. *Computational Statistics & Data Analysis*, vol 52, no 12, pp 5186–5201.

Kaber, D. B., Endsley, M. R. (1997). Out-of-the-loop performance problems and the use of intermediate levels of automation for improved control system functioning and safety. *Process Safety Progress*, vol 16, no 3, pp 126–131.

Kerschbaum, P., Lorenz, L., Bengler, K. (2015). A transforming steering wheel for highly automated cars. Paper presented at the Intelligent Vehicles Symposium (IV), IEEE.

Kircher, K., Larsson, A., Hultgren, J. A. (2014). Tactical driving behavior with different levels of automation. *IEEE Transactions on Intelligent Transportation Systems*, vol 15, no 1, pp 158–167.

Körber, M., Gold, C., Lechner, D., Bengler, K. (2016). The influence of age on the take-over of vehicle control in highly automated driving. *Transportation Research Part F: Traffic Psychology and Behaviour*, vol 39, pp 19–32.

Körber, M., Weißgerber, T., Kalb, L., Blaschke, C., Farid, M. (2015). Prediction of take-over time in highly automated driving by two psychometric tests. *Dyna*, vol 82, no 193, pp 195–201.

Lorenz, L., Kerschbaum, P., Schumann, J. (2014). Designing take over scenarios for automated driving: How does augmented reality support the driver to get back into the loop? In *Proceedings of the Human Factors and Ergonomics Society 58th Annual Meeting*, pp 1681–1685.

Louw, T., Kountouriotis, G., Carsten, O., Merat, N. (2015a). Driver inattention during vehicle automation: how does driver engagement affect resumption ff control? In *Proceedings of 4th International Conference on Driver Distraction and Inattention (DDI2015)*, Sydney, Australia: ARRB Group.

Louw, T., Merat, N., Jamson, H. (2015b). Engaging with highly automated driving: to be or not to be in the loop? Paper presented at the 8th International Driving Symposium on Human Factors in Driver Assessment, Training and Vehicle Design, Salt Lake City, Utah.

Lu, Z., de Winter, J. C. F. (2015). A review and framework of control authority transitions in automated driving. *Procedia Manufacturing*, vol 3, pp 2510–2517.

Melcher, V., Rauh, S., Diederichs, F., Widlroither, H., Bauer, W. (2015). Take-over requests for automated driving. *Procedia Manufacturing*, vol 3, pp 2867–2873.

Merat, N., Jamson, A. H., Lai, F. C., Carsten, O. (2012). Highly automated driving, secondary task performance, and driver state. *Human Factors*, vol 54, no 5, pp 762–771.

Merat, N., Jamson, A. H., Lai, F. C. H., Daly, M., Carsten, O. M. J. (2014). Transition to manual: Driver behaviour when resuming control from a highly automated vehicle. *Transportation Research Part F-Traffic Psychology and Behaviour*, vol 27, pp 274–282.

Michon, J. A. (1985). A critical view of driver behavior models: What do we know, what should we do? In L. Evans and R. C. Schwing (Eds.), *Human Behavior and Traffic Safety*. New York, NY: Plenum Press, pp 485–524.

Mok, B., Johns, M., Lee, K. J., Miller, D., Sirkin, D., Ive, P., Ju, W. (2015). Emergency, auto-mation off: Unstructured transition timing for distracted drivers of automated vehi-cles. Paper presented at the Intelligent Transportation Systems (ITSC), IEEE 18th International Conference.

Molloy, R., Parasuraman, R. (1996). Monitoring an automated system for a single failure: Vigilance and task complexity effects. *Human Factors*, vol 38, no 2, pp 311–322.

National Highway Traffic Safety Administration. (2013). Preliminary Statement of Policy Concerning Automated Vehicles, p 5

Naujoks, F., Mai, C., Nekum, A. (2014, 19–23 July). The effect of urgency of take-over requests during highly automated driving under distracted conditions. Paper presented at the 5th International Conference on Applied Human Factors and Ergonomics AHFE 2014, Krakow, Poland.

Naujoks, F., Nekum, A. (2014). Timing of in-vehicle advisory warnings based on cooperative perception. Paper presented at the Human Factors and Ergonomics Society Europe Chapter Annual Meeting.

Nilsson, J. (2014). Safe transitions to manual driving from faulty automated driving system. Doctoral thesis. Chalmers University of Technology, Gothenburg, Sweden.

Payre, W., Cestac, J., Delhomme, P. (2016). Fully automated driving: Impact of trust and prac-tice on manual control recovery. *Human Factors*, vol 58, no 2, pp 229–241.

Petermeijer, S. M., Abbink, D. A., Mulder, M., de Winter, J. C. (2015). The effect of hap-tic support systems on driver performance: A literature survey. *IEEE Transactions on Haptics*, vol 8, no 4, pp 467–479.

Petermeijer, S. M., de Winter, J. C. F., Bengler, K. J. (2016). Vibrotactile displays: A sur-vey with a view on highly automated driving. *IEEE Transactions on Intelligent Transportation Systems*, vol 17, no 4, pp 897–907.

Phillips, T. (2013). WPF Gauge 2013.5.27.1000. Retrieved from https://wpfgauge.codeplex. com.

Porter, J. M., Case, K., Marshall, R., Gyi, D., Oliver, R. S. N. (2004). 'Beyond Jack and Jill': Designing for individuals using HADRIAN. *International Journal of Industrial Ergonomics*, vol 33, no 3, pp 249–264.

Radlmayr, J., Gold, C., Lorenz, L., Farid, M., Bengler, K. (2014). How traffic situations and non-driving related tasks affect the take-over quality in highly automated driving. *Proceedings of the Human Factors and Ergonomics Society Annual Meeting*, vol 58, no 1, pp 2063–2067.

Robinson, T., Chan, E., Coelingh, E. (2010). Operating platoons on public motorways: An introduction to the sartre platooning programme. Paper presented at the 17th World Congress on Intelligent Transport Systems, Busan, South Korea.

SAE J3016. (2016). Taxonomy and definitions for terms related to driving automation systems for on-road motor vehicles, J3016_201609. SAE International.

Schömig, N., Hargutt, V., Neukum, A., Petermann-Stock, I., Othersen, I. (2015). The interac-tion between highly automated driving and the development of drowsiness. *Procedia Manufacturing*, vol 3, pp 6652–6659.

Scott, J. J., Gray, R. (2008). A comparison of tactile, visual, and auditory warnings for rear-end collision prevention in simulated driving. *Human Factors*, vol 50, no 2, pp 264–275.

Stanton, N. A., Dunoyer, A., Leatherland, A. (2011). Detection of new in-path targets by drivers using Stop & Go Adaptive Cruise Control. *Applied Ergonomics*, vol 42, no 4, pp 592–601.

Stanton, N. A., Marsden, P. (1996). From fly-by-wire to drive-by-wire: Safety implications of automation in vehicles. *Safety Science*, vol 24, no 1, pp 35–49.

Stanton, N. A., Young, M. S. (2005). Driver behaviour with adaptive cruise control. *Ergonomics*, vol 48, no 10, pp 1294–1313.

Stanton, N. A., Young, M. S., McCaulder, B. (1997). Drive-by-wire: The case of mental work-load and the ability of the driver to reclaim control. *Safety Science*, vol 27, no 2–3, pp 149–159.

Stanton, N. A., Young, M. S., Walker, G. H., Turner, H., and Randle, S. (2001). Automating the driver's control tasks. *International Journal of Cognitive Ergonomics*, vol 5, no 3, pp 221–236.

Strand, N., Nilsson, J., Karlsson, I. C. M., Nilsson, L. (2014). Semi-automated versus highly automated driving in critical situations caused by automation failures. *Transportation Research Part F-Traffic Psychology and Behaviour*, vol 27, pp 218–228.

Summala, H. (2000). Brake reaction times and driver behavior analysis. *Transportation Human Factors*, vol 2, no 3, pp 217–226.

Swaroop, D., Rajagopal, K. R. (2001). A review of constant time headway policy for automatic vehicle following. Paper presented at the IEEE Intelligent Transportation Systems Conference, Oakland, CA.

United Nations. (1968). Convention on Road Traffic, Vienna, 8 November 1968. Amendment 1. Retrieved from http://www.unece.org/fileadmin/DAM/trans/conventn/crt1968e.pdf.

van den Beukel, A. P., van der Voort, M. C. (2013). The influence of time-criticality on Situation Awareness when retrieving human control after automated driving. Paper presented at the Intelligent Transportation Systems (ITSC), IEEE 16th International Conference.

Verboven, S., Hubert, M. (2005). LIBRA: A MATLAB library for robust analysis. *Chemometrics and Intelligent Laboratory Systems*, vol 75, no 2, pp 127–136.

Walch, M., Lange, K., Baumann, M., Weber, M. (2015). Autonomous driving: investigating the feasibility of car-driver handover assistance. In *Proceedings of the 7th International Conference on Automotive User Interfaces and Interactive Vehicular Applications*, Nottingham, United Kingdom, pp 11–18.

Willemsen, D., Stuiver, A., Hogema, J. (2015). Automated driving functions giving control back to the driver: A simulator study on driver state dependent strategies. Paper presented at the 24th International Technical Conference on the Enhanced Safety of Vehicles (ESV), Gothenburg, Sweden.

Wolterink, W. K., Heijenk, G., Karagiannis, G. (2011). Automated merging in a Cooperative Adaptive Cruise Control (CACC) system. Paper Presented at the Fifth ERCIM Workshop on eMobility, Vilanova i la Geltrú, Catalonia, Spain.

Young, M. S., Stanton, N. A. (2002). Malleable attentional resources theory: A new explanation for the effects of mental underload on performance. *Human Factors*, vol 44, no 3, pp 365–375.

Young, M. S., Stanton, N. A. (2007). Back to the future: Brake reaction times for manual and automated vehicles. *Ergonomics*, vol 50, no 1, pp 46–58.

Zeeb, K., Buchner, A., Schrauf, M. (2015). What determines the take-over time? An integrated model approach of driver take-over after automated driving. *Accident Analysis and Prevention*, vol 78, pp 212–221.

Zeeb, K., Buchner, A., Schrauf, M. (2016). Is take-over time all that matters? The impact of visual-cognitive load on driver take-over quality after conditionally automated driving. *Accident Analysis and Prevention*, vol 92, pp 230–239.

5 Contrasting Simulated with On-Road Transitions of Control

Human Factors research into automated driving has been ongoing since the mid-1990s (Nilsson, 1995; Stanton and Marsden, 1996). As the motor industry advances towards highly automated driving, research conducted in driving simulators will become ever more important (Boer et al., 2015). As mentioned in Chapter 3, driving simulators have the advantage of allowing the evaluation of driver reactions to new technology within a virtual environment without the physical risk found on roads (Carsten and Jamson, 2011; De Winter et al., 2012; Flach et al., 2008; Nilsson, 1993; Stanton et al., 2001; Underwood et al., 2011). It is widely accepted that driving simulation offers a high degree of controllability and reproducibility and provides access to variables that are difficult to accurately determine in the real world (Godley et al., 2002), such as lane position and distance to roadway objects (van Winsum et al., 2000).

When evaluating the validity of a simulator, Blaauw (1982) and Santos et al. (2005) distinguished between two types of simulator validity: physical and behavioural validity. Physical validity refers to the level of correspondence between the physical layout, the configuration of the driver cabin, components and vehicle dynamics in the simulator and its real-world counterpart. Behavioural fidelity, or the correspondence in driver behaviour between the simulator and its on-road counterpart, is arguably the most important form of validity when it comes to the evaluation of a specific task (Blaauw, 1982). Behavioural fidelity can be further extended into absolute validity and relative validity. Absolute validity is obtained when the absolute size of an effect measured in a simulator is the same as the absolute effect measured in its on-road counterpart. Relative validity, on the other hand, describes how well the relative size or direction of an effect measured in the simulator corresponds to real driving (Blaauw, 1982; Kaptein et al., 1996).

Whilst there is plenty of research on the design of human–machine interfaces, driver errors and task load, very little of the research has demonstrated transfer from the simulated environment to the open road (Mayhew et al., 2011; Santos et al., 2005; Shechtman et al., 2009; Stanton et al., 2001, Stanton et al., 2011; Stanton and Salmon, 2009; Wang et al., 2010).

Most of the research showing how drivers interact with highly automated vehicles outside of simulators has taken place on closed test tracks (Albert et al., 2015; Llaneras et al., 2013; Stanton et al., 2011). There is only a minority of studies on highly automated driving performed on the road (Banks and Stanton, 2016; Endsley, 2017; Eriksson et al., 2017).

Other on-road studies on automation have investigated sub-systems such as adaptive cruise control (ACC) (Beggiato et al., 2015; Morando et al., 2016; Varotto et al.,

2015) and lane-keeping assistance systems (euroFOT, 2012; Ishida and Gayko, 2004; Stanton et al., 2001). This means that there is a paucity of research into the relative validity of driver behaviour in simulated highly automated vehicles. This lack of studies could be attributed to the costs and risks associated with non-professional drivers driving prototype vehicles (such as the Mercedes S/E-class and Tesla vehicles equipped with these features, for road testing [Mercedes-Benz, 2015; safecarnews.com, 2015; Tesla Motors, 2016]) Consequently, most research into human–automation interaction has been limited to simulators (for a review on control transitions in the simulator see Chapter 4) or closed test tracks (e.g. Albert et al., 2015; Llaneras et al., 2013; Stanton et al., 2011). Some disadvantages of testing on closed test tracks compared to on-road testing are the reduced complexity and the dissonance between driver behaviour on the track and in normal on-road driving, as well as the lack of other road users.

The purpose of the research reported in this chapter was to explore whether control transitions between automated driving and manual driving observed in a driving simulator study are similar to real-world driving. Chapter 4 found that drivers of manual vehicles (SAE Level 0) take approximately 1 second to respond to sudden events in traffic (Eriksson and Stanton, 2016). It was also found that drivers of 'function-specific automation' (ACC and assistive steering, SAE Level 1 and 2) took an additional 1.1–1.5 seconds to respond to a sudden automation failure, and that drivers of highly automated vehicles (SAE level 3) took on average 2.96 ± 1.96 seconds to respond to a control transition request leading up to a critical event, such as a stranded vehicle (Eriksson and Stanton, 2017). In contrast, Google (2015) reported that it takes their professional test drivers 0.84 seconds to respond to the automation failures of their autonomous (SAE Level 4/5) prototypes whilst driving on public roads, based on 272 discrete events. Moreover, the meta-analysis in Chapter 4 showed that the response-time varies with the lead time between the control transition request and a critical event. The reported lead times to the critical event at the point where the request from manual control was issued varied between 2 and 30 seconds and was 6.37 seconds on average.

This is somewhat problematic as the SAE guidelines for level 3 automation states that the driver 'Is receptive to a request to intervene and responds by performing dynamic driving task fallback in a timely manner' *(SAE J3016, 2016, p. 20)*. A decision to explore control transitions in non-urgent situations was made due to the lack of research into driver-paced transitions of control, which arguably is one of the more common use-cases for control transitions in highly automated driving when, for example, leaving a highway.

5.1   METHOD

This chapter is based on the results of a two-phase between-participant study. The first phase involved collecting times for control transitions within a simulated driving environment and the second phase collected the same data from the open road. The experimental design and procedure for each study are discussed in turn.

5.1.1   PHASE 1

The experimental set-up for the first phase of this chapter is identical to the set-up in Chapter 4 and is detailed in full in Sections 4.1.1 through 4.1.3.

5.1.2 PHASE 2

5.1.2.1 Participants

The second phase of the study was comprised of 12 participants (6 male, 6 female) between 20 and 49 years of age ($M = 32.33$, $SD = 10.98$) with a minimum of 1-year driving experience ($M = 14.58$, $SD = 11.13$). All participants in the on-road trial had undertaken extended driver training as a legal requirement for insurance purposes for the execution of phase 2 and therefore had previous experience with advanced driver assistance systems (ADAS), such as ACC or Lane Keeping Assist. Nevertheless, none of the drivers had previous experience with the Tesla Autopilot system. The second phase of the study was approved by the Southampton University ERGO ethics committee (RGO number 19151).

5.1.2.2 Equipment

Phase 2 of the study was conducted using a Tesla Model S P90 equipped with the Autopilot software feature, which enables short periods of hands- and feet-free driving on motorways as longitudinal and lateral control becomes automated (Figure 5.1). Drivers were reminded that they were ultimately responsible for safe vehicle operation and were not actively encouraged to remove their hands from the wheel at any point during the study. To ensure consistency between the two experiments, an iPad was mounted next to the instrument cluster running the application 'Duet Display'. This enables the iPad to act as a secondary monitor, displaying the same type of take-over request visual feedback and auditory messages as in the simulator trial. The take-over requests were reset by the experimenter, sat in the rear of the vehicle, in a Wizard-of-Oz fashion (Dahlbäck et al., 1993). To capture the control transitions a Video VBox Pro from Racelogic was used. Participants were invited to drive along public roads and highways within Warwickshire, United Kingdom (B4100, M40 and M42). They were asked to adhere to national speed limits at all times and to keep lane changes to a minimum. Data recording of take-over request response-times took place on the M40 and M42 where speed was limited to 70 mph.

FIGURE 5.1 In-vehicle view from a Tesla Model S. (Image credit: Daniël Heikoop.)

5.2 DEPENDENT VARIABLES

- Reaction time to the control transition request was recorded from the onset of the take-over request until the driver switched mode in the simulated driving condition. In the on-road driving condition the reaction time was measured from the first frame visible in the Wizard-of-Oz interface until the first frame where the mode-indicator in the Tesla instrument cluster had changed, indicating a change in automated driving mode. All transitions were measured in milliseconds.
- A technology acceptance questionnaire (Van Der Laan et al., 1997) was used to measure the usefulness and satisfaction of the automation in the simulated and on-road drive. The usefulness score was determined across the following items on a semantic-differential five-point scale: useful–useless, bad–good, effective–superfluous, assisting–worthless, and raising alertness–sleep-inducing. The satisfaction score was determined by four items (2, 4, 6, 8), the usefulness score by the remaining five.
- The NASA raw task load index (TLX) was used to evaluate the perceived workload per condition (Byers et al., 1989; Hart and Staveland, 1988).

5.3 PROCEDURE

Upon providing informed consent, participants in both phases of the study were provided with additional information about the highly automated driving feature they would be driving with. They were told that the Tesla system could be overridden via the steering wheel, throttle or brake pedals and through a touchscreen interface in the simulator. Participants were reminded that they were responsible for the safe operation of the vehicle at all times, regardless of its mode (manual or automated) in accordance with recent amendments to the Vienna Convention on Road Traffic (United Nations, 1968). Participants were told that the system may prompt them to either resume or relinquish control of the vehicle during the drive and that they should adhere to the instruction only when they felt it was safe to do so. This was intended to reduce the pressure on participants to respond immediately and to reinforce the idea that they were ultimately responsible for safe vehicle operation. For the 12 participants involved in the on-road study, additional instructions were given to ensure they remained aware of the vehicle's internal human–machine interface (specifically about the state of the Autopilot) in an effort to maintain the safety of the vehicle driver and passengers in case of Autopilot malfunction, or failure to engage the Autopilot due to, for example, missing lane markings. To support them in doing this, a qualified safety driver was present in the passenger seat at all times, ready to prompt the drivers to take back control if the need to regain control arose, or to press the emergency stop button in the Tesla centre display should the driver be unresponsive to prompts by the safety driver.

In both phases of the study, control transition requests were presented as both a visual cue (Figure 5.2) and an auditory message in the form of a computer-generated, female voice stating 'please resume control' or 'automation available'. The interval in which these requests were issued ranged from 30 to 45 seconds, thus allowing for approximately 24 control transitions, half of which were to manual during the approximately 20-minute drive on the M40 and M42.

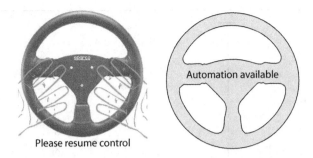

FIGURE 5.2 The left-hand side, the instrument cluster showing a take-over request. The visual take-over request was coupled with a computer-generated voice message stating 'please resume control'. On the right-hand side is a control transitions request to automated vehicle control presented in the instrument cluster, coupled with a computer-generated voice message stating 'automation available'.

In the simulated driving condition, the human–machine interface used to switch mode was located in the centre display, running on a windows tablet, consisting of two buttons used to engage or disengage the automated driving feature. The automated driving system in the simulator was set to disengage only when the mode-switching buttons were pressed to allow for consistent control transitions.

In the on-road driving condition, the automated driving feature was engaged by a 'double pull' on a control stork on the left side of the steering wheel, below the indicator stork. To disengage the automated driving feature, the driver could either depress the brake to disengage both the ACC and lateral control, apply a steering input to disengage the lateral control only or press the control stork forwards to disengage the ACC and lateral control. Reaction times to the control transition request were recorded for each participant. In phase 1, reaction time was recorded from the onset of stimuli until the driver completed the requested action.

In phase 2, reaction times to control transition requests were captured through a Racelogic video VBOX Pro at a rate of 30 Hz and manually coded in a frame-by-frame fashion using video editing software. The response-time in the on-road study was based upon the mode-indicator in the Tesla instrument cluster switching mode after the control transition request was displayed on the iPad display. At the end of each drive, participants were asked to fill out the NASA-TLX (Hart and Staveland, 1988) and Technology Acceptance Scale (Van Der Laan et al., 1997) with respect to the control transition process.

5.4 ANALYSIS

The median take-over reaction time values for each participant were calculated (Baayen and Milin, 2010), after which Wilcoxon rank sum tests were computed to analyse response-time and TLX data. The box plots in Figure 5.3 were adjusted to accommodate the log-normal distribution of the take-over response-time data (Hubert and Vandervieren, 2008).

To enable correlation analysis of take-over response-times between the two groups of participants, drivers from the simulated drive were matched with drivers from the

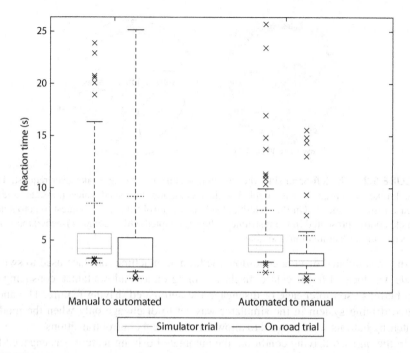

FIGURE 5.3 Control transition times from manual to automated control and from automated to manual control in on-road and simulated driving condition.

on-road driving scenario based on gender, age and driving experience, as shown in Table 5.1. An uneven sample size was still present after participant matching, with fewer transitions recorded in the on-road condition. Therefore, a randomised removal of observations on a participant-by-participant basis was conducted to ensure an equal number of observations for each participant pair. After data reduction, the take-over response-times for each task condition were added to a single vector and sorted in ascending order which, according to Ryan and Joiner (1976), enables a comparison of the two distributions to be made. As this dataset is used in more than one chapter in this book, the significance (alpha) level in this chapter is set to 0.01.

5.5   RESULTS

The results showed significant differences between on-road and simulated driving when relinquishing control to the vehicle automation ($Z=-6.120$, $p<0.01$). On average, a 1-second increase in response-time was found (see Table 5.2, Figure 5.3) when relinquishing control to the vehicle automation in the simulated road condition. It took drivers approximately 3.18 ± 2.83 seconds to relinquish control in the on-road condition, whilst it took 4.20 ± 1.96 seconds to relinquish control in the simulated driving condition.

Moreover, a significant decrease in take-over request reaction time of approximately 1.5 seconds was found in the on-road driving condition, in comparison to the simulated driving condition, when resuming control from the automation

TABLE 5.1

Participants Matched on Age and Gender from the On-Road and Simulator Experiments

		On Road		Simulator	
Participant	Gender	Age	Driving Experience (Years)	Age	Driving Experience (Years)
1	Male	20	3	20	1
2	Male	24	3	24	7
3	Female	29	12	29	6
4	Male	59	43	52	35
5	Female	32	15	28	10
6	Male	26	9	24	6
7	Male	28	11	28	6
8	Female	30	12	28	10
9	Female	32	14	40	22
10	Female	49	28	50	33
11	Female	26	8	27	10
12	Male	33	17	34	17
Mean		32.33	14.58	32	13.58
SD		10.98	11.13	10.20	10.99

The participants are ordered based on participant number for the on-road trial.

$(Z = -10.403, p < 0.01)$. It took drivers approximately 3.08 ± 1.16 seconds to resume control from the automation in the on-road driving condition and 4.56 ± 1.63 seconds to resume control in the simulated driving condition. The results from the correlation analysis of the sorted response-time data for the transition to automated driving from manual driving showed a strong positive correlation (Pearson's $r = 0.96$, $p < 0.0001$, calculated power $= 1.0$), as illustrated in Figure 5.4.

TABLE 5.2

Control Transition Times (seconds) between Manual and Automated and Automated and Manual Control in On-Road and Simulator Conditions

	To Automated		To Manual	
	Simulator	On Road	Simulator	On Road
Median	4.20	3.18	4.56	3.08
IQR	1.96	2.83	1.63	1.16
Min	2.82	1.33	1.97	1.21
Max	23.88	25.16	25.75	15.41

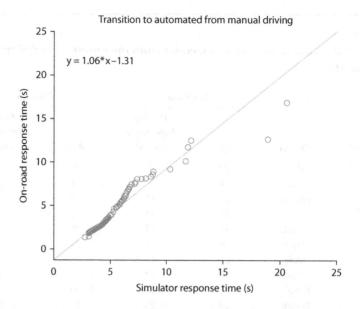

FIGURE 5.4 Scatter plot of control transitions from manual to automated vehicle control. The X axis shows the driver response-time in the simulator, the Y-axis shows the driver response-time in the on-road condition.

The correlation analysis of the control transition time from automated to manual control showed a significant positive relationship (Pearson's $r=0.97$, $p<0.0001$, calculated power = 1.0) between the two distributions as shown in Figure 5.5.

Subjective workload scores collected through the NASA-TLX subscales (Byers et al., 1989; Hart and Staveland, 1988) at the end of each driving condition showed no significant correlations on the workload subscales nor overall workload for the matched sample. A comparison of the two conditions showed no significant differences on any of the subscales or overall workload (Table 5.3). Overall, there was little difference in workload, and the median workload in both conditions was approaching the halfway point on the scale (Figure 5.6), implying relatively low workload.

The Van Der Laan Technology Acceptance Scale yielded no significant correlations on the matched samples as shown in Table 5.3. Moreover, no significant differences in automation usefulness ($Z=0.491$ $p>0.05$, $r=0.07$) and automation satisfaction ($Z=1.316$ $p>0.05$, $r=0.21$) between the on-road and simulated driving conditions could be found.

5.6 DISCUSSION

As shown in Chapter 4, most research into control transitions in highly automated driving has been undertaken in driving simulators. The studies outside of simulators tend to be limited to closed test tracks (Albert et al., 2015; Llaneras et al., 2013) or to sub-systems of highly automated driving (Beggiato et al., 2015; euroFOT, 2012; Ishida and Gayko, 2004; Morando et al., 2016; Varotto et al., 2015). One study by

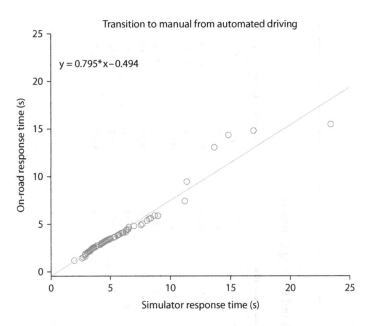

FIGURE 5.5 Scatter plot of control transitions from automated to manual vehicle control. The X axis shows the driver response-time in the simulator, the Y-axis shows the driver response-time in the on-road condition.

Banks and Stanton (2016) explored the interaction with automated vehicles on the open road, but links to performance on similar tasks in simulated environments were not made. Another study by Endsley (2017) explored the experience and behaviour after longitudinal use of a Tesla Model S (with one participant). To further the understanding of the validity of driving simulators in highly automated driving, multiple control transitions from manual driving to automated driving and vice versa were compared for both simulated and on-road driving environments. The reason for the unusually high frequency of transitions of control compared with their occurrence in contemporary literature was to reduce the impact of novelty effects, to enable the capturing of inter- and intra-individual differences to get an appreciation for the wide range of transition times and to compress experience with the system as previously done by Stanton et al. (2001). Moreover, it was argued in Chapter 4 that the type of control transition utilised in this chapter is 'non-urgent' and such transitions are likely to be commonplace on public roads when SAE level 3 systems are limited to certain operational constraints (SAE J3016, 2016).

The results show that drivers in the on-road driving condition took on average 3.08 seconds. This is marginally longer compared to the average 2.96 seconds control resumption time for drivers who are required to resume control within a limited time frame (e.g. 7 seconds as in Gold et al. [2013]). It has previously been shown that permitting drivers to self-regulate the use of in-vehicle technologies on a tactical level tends to maintain optimal workload and safer driving performance (Cooper et al., 2009; Eriksson et al., 2014; Eriksson and Stanton, 2017; Kircher et al., 2016;

TABLE 5.3
NASA-TLX and Technology Acceptance Scores of the On-road and Simulated Driving Conditions

	On Road Median (SD)	Simulator Median (SD)	Rank Sum Test			Pearson's Correlation	
			Z	p	r	p	r
Mental demand	8 (7)	7.5 (10)	−0.603	0.53	0.09	0.60	−0.16
Physical demand	2.5 (4.5)	4 (5)	1.395	0.16	0.23	0.71	−0.12
Temporal demand	4.5 (5)	3 (6)	−0.415	0.68	0.06	0.81	−0.08
Performance	5 (2.5)	6 (7)	0.872	0.38	0.14	0.81	−0.08
Effort	5.5 (5)	5 (9)	0.206	0.84	0.03	0.66	0.14
Frustration	4.5 (4)	4 (9)	0.269	0.79	0.04	0.25	−0.35
Overall workload	4.75 (4.2)	5.16 (7.33)	0.755	0.45	0.12	0.92	−0.03
Usefulness	1.1 (0.6)	1 (1)	0.491	0.62	0.07	0.64	−0.14
Satisfaction	1 (0.87)	0.5 (1.69)	1.316	0.18	0.21	0.43	−0.25

The effect size was calculated as $r = |z|/\sqrt{(N1+N2)}$.

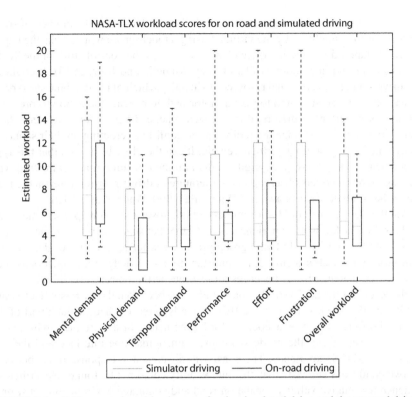

FIGURE 5.6   Self-reported workload scores for the simulated drive and the on-road drive.

Young and Stanton, 2007). Moreover, the results show that there is a long tail in the resumption-time distribution, indicating that some drivers take up to 15 seconds to resume control.

This shows that the design of the take-over process should not focus on average resumption times, but rather use the 5th–95th percentile user, as argued in Chapter 4 and is common practice in anthropometrics (Eriksson and Stanton, 2017; Porter et al., 2004). The results also show that drivers were generally faster in the execu- tion of control transitions in the on-road driving scenario for transitions to both automated and manual control, compared to the simulated driving scenario. Similar effects have been observed by Wang et al. (2010) and Kurokawa and Wierwille (1990), where drivers produced faster responses in on-road conditions than in the simulator for in-vehicle interaction tasks.

This difference could be partially explained by the perception of greater risk in the on-road condition (Carsten and Jamson, 2011; De Winter et al., 2012; Flach et al., 2008; Underwood et al., 2011). Moreover, the differences between mode-switch- ing human–machine interface in the Tesla (control stork next to the steering wheel) and the simulator (touchscreen in centre console) could account for the increased response-time in the simulator part of this study. Drivers have been found to have significantly higher eyes-off-road time when engaged with in-vehicle systems with high visual demands, and as driving is a visually demanding task, this can have large effects on driving performance (Jæger et al., 2008). This increase in eyes-off-road could be further amplified by the virtue of using a touchscreen which lacks the haptic nature of standard vehicle interface elements, such as the control stork in the Tesla, that enables blind interaction whilst driving (Rümelin and Butz, 2013). It could be that drivers had to divert visual resources to identify which of the two buttons to press to reach the desired state and to plan a motor-path to execute the action to press the button. For experienced drivers, this is a well-practiced behaviour, executed whenever a driver needs to change the radio station, confirm a rerouting on their satnav or change the heating settings of their vehicle. It can therefore be argued that this type of interaction should have a negligible effect on the transition times compared with the magnitude of effects observed in the literature of driver reaction time on the road and in the simulator (Kurokawa and Wierwille, 1990; Wang et al., 2010).

Another factor that could have influenced take-over request response-time could be the different levels of experience with ADAS between the two samples, where drivers in the experienced (Tesla) group produced faster reaction times due to their familiarity with such systems. It is important to acknowledge these factors as they may have contributed to the faster response-times in the on-road condition.

However, as effects of a similar magnitude have been found previously (Kurokawa and Wierwille, 1990; Wang et al., 2010), it stands to reason that the main part of the observed difference can be accounted for by the nature of the driving environment, and to a lesser extent, the mode-switching human–machine interface and the differences in ADAS experience. The consistent difference in response-times between the two conditions suggests that absolute validity could not be found. Nevertheless, evidence for relative validity for the on-road and simulated environments is strong, as the correlation analysis showed a strong positive similarity in the distributions for both types of control transitions. The lack of correlations in workload and technology acceptance scores could be explained by the subjective nature of the questionnaire, combined with a relatively small sample in a between-group study-design. It is worth noting that the workload and technology acceptance scores are not absolute scores, and they therefore need to be looked upon with some caution due to the relatively small sample of two independent groups.

Despite the lack of correlation between the on-road and simulated drive on the subjective questionnaires, the lack of differences in workload and technology acceptance scores indicates that the driving conditions had no measurable effect on perceived workload, usefulness and satisfaction. It is therefore argued that the relative fidelity of the simulator can be established with regard to human–automation interaction and, in particular, control transitions (Blaauw, 1982; Godley et al., 2002). These results support Stanton et al. (2001), who found that driver performance on secondary tasks was highly correlated when performed on the road and in the simulator during manual driving.

Consequently, the results obtained in this study lend validity to previous research into control transitions in automated vehicles carried out in simulated environments. In light of these results, researchers may have more confidence when using simulators as a primary tool for research on human–automation interaction (Stanton et al., 2001). This observation permits the exploration of phenomena related to automated vehicles in a reproducible, deterministic and completely observable environment (Russel and Norvig, 2009), and facilitates the collection of data that would otherwise be difficult to obtain in road vehicles (Godley et al., 2002; Santos et al., 2005; van Winsum et al., 2000). These findings show that driving simulators are legitimate tools for researching vehicle automation (Boer et al., 2015).

5.7 CONCLUSIONS

In this chapter, the validity of human–automation interaction in highly automated vehicles in driving simulators was assessed. Absolute validity could not be established due to the shorter transition times observed in the on-road driving condition. It was found that, on average, drivers take an additional second to transfer control to the automation in the simulated drive, and an additional 1.4 seconds to resume control in the simulated drive compared to on-road. Moreover, it was found that drivers in the on-road driving condition were marginally faster than what has been found in previous literature when drivers have resumed control under time pressure. Despite these similarities, it was also shown that there is a long tail of the distribution of control resumption times and that these drivers will have to be accommodated to ensure the safe use of automation. Nevertheless, the results also showed that there was a strong positive correlation for transition time in the on-road and simulated driving conditions. In light of these results, it is argued that there is a strong indication of relative validity for research conducted in simulators. Despite the lack of significant correlations, relative validity is further supported by the similarities in workload and technology acceptance scores of the drivers in the simulated and on-road driving conditions.

Consequently, in this study, it is argued that the driving simulator is a valid research tool for the exploration of human–automation interaction and in particular the transfer of control between driver and automation. In conclusion, medium-fidelity, fixed-based driving simulation is a safe and cost-effective method for assessing human–automation interaction, and in particular control transitions in highly automated driving.

5.8 FUTURE DIRECTIONS

Chapter 5 provided an in-depth analysis of the effects of automated driving on driving performance after resuming manual control from the automated vehicle in the driving simulator, as this chapter established the validity of the algorithms described in Chapter 3 and used in Chapter 4. It is anticipated that any after-effects will be negligible when drivers are able to self-pace the transition process, contrary to contemporary literature utilising system-paced control transitions.

REFERENCES

Albert, M., Lange, A., Schmidt, A., Wimmer, M., Bengler, K. (2015). Automated driving–Assessment of interaction concepts under real driving conditions. *Procedia Manufacturing*, vol 3, pp 2832–2839.

Baayen, R. H., Milin, P. (2010). Analyzing reaction times. *International Journal of Psychological Research*, vol 3, no 2, pp 12–28.

Banks, V. A., Stanton, N. A. (2016). Keep the driver in control: Automating automobiles of the future. *Applied Ergonomics*, vol 53, Pt B, pp 389–395.

Beggiato, M., Pereira, M., Petzoldt, T., Krems, J. (2015). Learning and development of trust, acceptance and the mental model of ACC. A longitudinal on-road study. *Transportation Research Part F-Traffic Psychology and Behaviour*, vol 35, pp 75–84.

Blaauw, G. J. (1982). Driving experience and task demands in simulator and instrumented car: A validation study. *Human Factors: The Journal of the Human Factors and Ergonomics Society*, vol 24, no 4, pp 473–486.

Boer, E. R., Penna, M. D., Utz, H., Pedersen, L., Sierhuis, M. (2015, 16–18 September). The role of driving simulators in evaluating autonomous vehicles. Paper Presented at the Driving Simulation Conference, Max Planck Institute for Biological Cybernetics, Tübingen, Germany.

Byers, J. C., Bittner, A., Hill, S. (1989). Traditional and raw task load index (TLX) correlations: Are paired comparisons necessary. In A. Mital (Ed.), *Advances in Industrial Ergonomics and Safety I*. London: Taylor & Francis, pp 481–485.

Carsten, O., Jamson, A. H. (2011). Driving simulators as research tools in traffic psychology. In B. Porter (Ed.), *Handbook of Traffic Psychology* vol 1. London: Academic Press, pp 87–96.

Cooper, J. M., Vladisavljevic, I., Medeiros-Ward, N., Martin, P. T., Strayer, D. L. (2009). An investigation of driver distraction near the tipping point of traffic flow stability. *Human Factors*, vol 51, no 2, pp 261–268.

Dahlbäck, N., Jönsson, A., and Ahrenberg, L. (1993). Wizard of Oz studies: Why and how. Paper Presented at the 1st International Conference on Intelligent User Interfaces, Orlando, FL, January 04–07, 1993.

De Winter, J. C. F., van Leeuwen, P., Happee, R. (2012). Advantages and disadvantages of driving simulators: A discussion. Paper Presented at the Measuring Behavior Conference, Utrecht, the Netherlands, August 28–31, 2012.

Endsley, M. R. (2017). Autonomous driving systems: A preliminary naturalistic study of the Tesla Model S. *Journal of Cognitive Engineering and Decision Making*, vol 11, no 3, pp 225–238.

Eriksson, A., Banks, V. A., Stanton, N. A. (2017). Transition to manual: Comparing simulator with on-road control transitions. *Accident Analysis & Prevention*, vol 102, pp 227–234.

Eriksson, A., Lindström, A., Seward, A., Seward, A., Kircher, K. (2014). Can user-paced, menu-free spoken language interfaces improve dual task handling while driving? In M. Kurosu (Ed.), *Human-Computer Interaction. Advanced Interaction Modalities and Techniques. HCI 2014 Lecture Notes in Computer Science* (vol 8511). Cham, Switzerland: Springer, pp 394–405.

Eriksson, A., Stanton, N. A. (2017). Takeover time in highly automated vehicles: Noncritical transitions to and from manual control. *Human Factors*, vol 59, no 4, pp 689–705.

euroFOT. (2012). euroFOT – Bringing intelligent vehicles to the road. Retrieved from http://www.eurofot-ip.eu/.

Flach, J., Dekker, S., Stappers, P. J. (2008). Playing twenty questions with nature (the surprise version): Reflections on the dynamics of experience. *Theoretical Issues in Ergonomics Science*, vol 9, no 2, pp 125–154.

Godley, S. T., Triggs, T. J., Fildes, B. N. (2002). Driving simulator validation for speed research. *Accident Analysis Prevention*, vol 34, no 5, pp 589–600.

Gold, C., Damböck, D., Lorenz, L., Bengler, K. (2013). "Take over!" How long does it take to get the driver back into the loop? In *Proceedings of the Human Factors and Ergonomics Society Annual Meeting*, vol 57, no 1, pp 1938–1942.

Google. (2015). Google self-driving car testing report on disengagements of autonomous mode. Retrieved from https://static.googleusercontent.com/media/www.google.com/en//selfdrivingcar/files/reports/report-annual-15.pdf.

Hart, S. G., Staveland, L. E. (1988). Development of NASA-TLX (Task Load Index): Results of empirical and theoretical research. *Advances in Psychology*, vol 52, pp 139–183.

Hubert, M. Vandervieren, E. (2008). An adjusted boxplot for skewed distributions. *Computational Statistics & Data Analysis*, vol 52, no 12, pp 5186–5201.

Ishida, S.,Gayko, J. E.(2004). Development, evaluation and introduction of a lane keeping assistance system. Paper Presented at the Intelligent Vehicles Symposium, 2004 IEEE, Parma, Italy.

Jæger, M. G., Skov, M. B., Thomassen, N. G. (2008). You can touch, but you can't look: Interacting with in-vehicle systems. In *Proceedings of the SIGCHI Conference on Human Factors in Computing Systems*, Florence, Italy, pp 1139–1148.

Kaptein, N., Theeuwes, J., Van Der Horst, R. (1996). Driving simulator validity: Some considerations. *Transportation Research Record: Journal of the Transportation Research Board*, vol 1550, pp 30–36.

Kircher, K., Eriksson, O., Forsman, Å., Vadeby, A., Ahlstrom, C. (2016). Design and analysis of semi-controlled studies. *Transportation Research Part F: Traffic Psychology and Behaviour*, vol 46 Pt B, pp 404–412.

Kurokawa, K., Wierwille, W. W. (1990). Validation of a driving simulation facility for instrument panel task performance. In *Proceedings of the Human Factors and Ergonomics Society Annual Meeting*, October 1990, Santa Monica, CA, SAGE Publications.

Llaneras, R. E., Salinger, J., Green, C. A. (2013). Human factors issues associated with limited ability autonomous driving systems: Drivers' allocation of visual attention to the forward roadway. In *Proceedings of the 7th International Driving Symposium on Human Factors in Driver Assessment, Training and Vehicle Design*, Bolton Landing, New York, NY, pp 92–98.

Mayhew, D. R., Simpson, H. M., Wood, K. M., Lonero, L., Clinton, K. M., Johnson, A. G. (2011). On-road and simulated driving: Concurrent and discriminant validation. *Journal of Safety Research*, vol 42, no 4, pp 267–275.

Mercedes-Benz. (2015). Intelligent drive next level as part of driving assistance package. https://www.mercedes-benz.com/en/mercedes-benz/innovation/with-intelligent-drive-more-comfort-in-road-traffic/.

Morando, A., Victor, T., Dozza, M. (2016). Drivers anticipate lead-vehicle conflicts during automated longitudinal control: Sensory cues capture driver attention and promote appropriate and timely responses. *Accident Analysis and Prevention*, vol 97, pp 206–219.

Nilsson, L. (1993). Behavioural research in an advanced driving simulator-experiences of the VTI system. In *Proceedings of the Human Factors and Ergonomics Society Annual Meeting*, vol 37, no 9, pp 612–616.

Nilsson, L. (1995). Safety effects of adaptive cruise control in critical traffic situations. *Paper Presented at the the Second World Congress on Intelligent Transport Systems: 'Steps Forward'*, Yokohama, Japan.

Porter, J. M., Case, K., Marshall, R., Gyi, D., Oliver, R. S. N. (2004). 'Beyond Jack and Jil l': Designing for individuals using HADRIAN. *International Journal of Industrial Ergonomics*, vol 33, no 3, pp 249–264.

Rümelin, S., Butz, A. (2013). How to make large touch screens usable while driving. In *Proceedings of the 5th International Conference on Automotive User Interfaces and Interactive Vehicular Applications*, Eindhoven, the Netherlands, pp 48–55.

Russel, S., Norvig, P. (2009). *Artificial Intelligence: A Modern Approach* (3rd Ed.). Upper Saddle River, NJ: Pearson Education.

Ryan, T., Joiner, B. (1976). Normal probability plots and tests for normality. Minitab Statistical Software: Technical Reports. The Pennsylvania State University, State College, PA. Available from MINITAB.

SAE J3016. (2016). Taxonomy and definitions for terms related to driving automation systems for on-road motor vehicles, *J3016_201609*. SAE International.

safecarnews.com. (2015). Intelligent Drive Concept for New Mercedes-Benz GLC. From http://safecarnews.com/intelligent-drive-concept-for-new-mercedes-benz-glc_ju6145/

Santos, J., Merat, N., Mouta, S., Brookhuis, K., de Waard, D. (2005). The interaction between driving and in-vehicle information systems: Comparison of results from laboratory, simulator and real-world studies. *Transportation Research Part F-Traffic Psychology and Behaviour*, vol 8, no 2, pp 135–146.

Shechtman, O., Classen, S., Awadzi, K., Mann, W. (2009). Comparison of driving errors between on-the-road and simulated driving assessment: A validation study. *Traffic Injury Prevention*, vol 10, no 4, pp 379–385.

Stanton, N. A., Dunoyer, A.,Leatherland, A.(2011). Detection of new in-path targets by drivers using Stop & Go Adaptive Cruise Control. *Applied Ergonomics*, vol 42, no 4, pp 592–601.

Stanton, N. A., Marsden, P. (1996). From fly-by-wire to drive-by-wire: Safety implications of automation in vehicles. *Safety Science*, vol 24, no 1, pp 35–49.

Stanton, N. A., Salmon, P. M. (2009). Human error taxonomies applied to driving: A generic driver error taxonomy and its implications for intelligent transport systems. *Safety Science*, vol 47, no 2, pp 227–237.

Stanton, N. A., Young, M. S., Walker, G. H., Turner, H., Randle, S. (2001). Automating the driver's control tasks. *International Journal of Cognitive Ergonomics*, vol 5, no 3, pp 221–236.

Tesla Motors. (2016). Model S software version 7.0. Retrieved from https://www.teslamotors.com/presskit/autopilot

Underwood, G., Crundall, D., Chapman, P. (2011). Driving simulator validation with hazard perception. *Transportation Research Part F-Traffic Psychology and Behaviour*, vol 14, no 6, pp 435–446.

United Nations. (1968). Convention on Road Traffic, Vienna, 8 November 1968. Amendment 1. Retrieved from http://www.unece.org/fileadmin/DAM/trans/conventn/crt1968e.pdf.

Van Der Laan, J. D., Heino, A., De Waard, D. (1997). A simple procedure for the assessment of acceptance of advanced transport telematics. *Transportation Research Part C: Emerging Technologies*, vol 5, no 1, pp 1–10.

van Winsum, W., Brookhuis, K. A., de Waard, D. (2000). A comparison of different ways to approximate time-to-line crossing (TLC) during car driving. *Accident Analysis & Prevention*, vol 32, no 1, pp 47–56.

Varotto, S. F., Hoogendoorn, R. G., van Arem, B., Hoogendoorn, S. P. (2015). Empirical longitudinal driving behavior in authority transitions between adaptive cruise control and manual driving. *Transportation Research Record*, vol 2489, pp 105–114.

Wang, Y., Mehler, B., Reimer, B., Lammers, V., D'Ambrosio, L. A., Coughlin, J. F. (2010). The validity of driving simulation for assessing differences between in-vehicle informational interfaces: A comparison with field testing. *Ergonomics*, vol 53, no 3, pp 404–420.

Young, M. S., Stanton, N. A. (2007). Back to the future: Brake reaction times for manual and automated vehicles. *Ergonomics*, vol 50, no 1, pp 46–58.

6 After-Effects of Driver-Paced Transitions of Control

Whilst automated vehicles show promise in reducing road accidents (Eriksson et al., 2017; World Health Organization, 2015), they are, in their current form, no panacea for road safety (Eriksson and Stanton, 2017b; Gold et al., 2013; 2016; Kalra and Paddock, 2016). As drivers engage a contemporary automated driving system, they are decoupled from the operational and tactical levels of control, leaving only high-level strategic goals to be dealt with by the driver (Bye et al., 1999; Michon, 1985) whilst still being expected to resume control when the vehicle reaches the limits of its Operational Design Domain (ODD, the ODD may include geographic, roadway, environmental, traffic, speed and/or temporal limitations of automated driving availability; SAE J3016, 2016) or when a system failure, or sudden, unexpected event forces a transition back to manual control (SAE J3016, 2016). This intermediate form of automation has been deemed hazardous as drivers are required to monitor the automation and be able to regain control of the vehicle at all times (Casner and Schooler, 2015; Seppelt and Victor, 2016; Stanton, 2015). This is a form of 'driver-initiated automation', where the driver is in control regardless of whether the system is engaged or disengaged (Banks and Stanton, 2015, 2016; Lu et al., 2016; SAE J3016, 2016), contrary to 'system-initiated automation', where the controlling agent, be it driver or the automated driving system, is in control as determined by artificial intelligence (AI) (Gordon et al., 2017) in a Men Are Better At—Machines Are Better At (MABA–MABA) fashion (Dekker and Woods, 2002). The intermittent transitions of control and the sharing of task-relevant information between driver and driving automation can be described in terms of distributed cognition (DCOG) (Hollan et al., 2000; Hutchins, 1995). DCOG offers a paradigm shift in describing how a human interacts with other humans, artefacts and artificial agents, describing it as a 'system' where cognition, knowledge and mental models (i.e. Common Ground) are distributed between agents in the Joint Cognitive System, henceforth referred to as *system* (Hollnagel and Woods, 2005). Chapter 2 stated that in such a *system*, coordination and communication between system entities is of the utmost importance (Christoffersen and Woods, 2002; Eriksson and Stanton, 2017a; Hutchins, 1995; Stanton, 2014).

The functioning of a cognitive *system* has been described by Hollnagel and Woods (2005) in the COntextual COntrol Model (COCOM) as a cybernetics-inspired tracking loop. The COCOM model describes how the control of a system can be lost and regained, and how different levels of control influence performance. As time progresses in a task such as driving, dynamic shifting between four different control

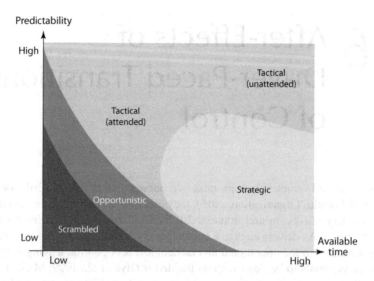

FIGURE 6.1 The relationship between Hollnagel and Woods (2005) control levels and available time and predictability of a situation.

modes can be made, dependent on the time horizon, and the predictability of the situation (Figure 6.1).

Hollnagel and Woods (2005) describe the four control modes in the following way:

- In the *scrambled* control mode, control actions are selected at random in a trial-and-error fashion, often urgently.
- In the *opportunistic* control mode, the next action to be carried out is determined by the salient features of the current context, such as a dominating part of the human–machine interface, with limited forward planning and anticipation. This control mode relies on internal heuristics and may be inefficient compared with higher levels of control.

  The driver dependence on the human–machine interface in this control mode is important, and it stands to reason that if the information in the human–machine interface can be tailored to both situational and temporal factors (Eriksson et al., 2015; Eriksson and Stanton, 2017a), the driver may transfer into a higher mode. For this to be successful, it requires the feedback from the human–machine interface to be designed in accordance to the Gricean maxims (presented in Chapter 2) in a way that is relevant, clearly conveyed in the right modality without oversaturating the information channels.

- In the *tactical* control mode, the next action is usually pre-planned, as the operator has a longer time horizon, thus enabling the use of rules and procedures to carry out actions. In the *tactical* control mode, the operator is still heavily influenced by the immediacy of the situation and will therefore still be influenced by the interface to some extent (Stanton et al., 2001).

Consequently, temporally and contextually relevant feedback may be used to facilitate maintaining or achieving higher levels of control, even as the driver reaches a tactical level of control. As shown in Figure 6.1, the tactical control mode can be divided into two sub-levels, namely attended and unattended. This division is made as when the level of familiarity (i.e. predictability) and available time in a situation is high; operators will go into unattended control which may reduce the thoroughness of a task. An example would be driving on a familiar route, such as the one going to the office but with another destination in mind, causing the driver to turn off towards the wrong destination (this could be mitigated through the use of a satnav). The distinction between the attended and unattended level of tactical control is described as the former being meticulously and carefully executed; whereas in the latter, drivers know what to do but do not bother to follow it in detail, thus increasing the likelihood of failure (Hollnagel and Woods, 2005).

- In the highest level control mode, the *strategic* control mode, the time horizon is longer (Figures 6.1 and 6.2), which enables long-term planning and anticipation. Operators in this control mode have evaluated the relationship between cause and effect more precisely, and will, therefore, have more overall control of the situation (Stanton et al., 2001).

There is a linear progression through the different control states as shown in Figure 6.2, meaning that once control is lost and an operator has reached the scrambled control state, the operator must progress through the intermediate stages before high-level strategic control can be reached. However, there is a 'shortcut' between

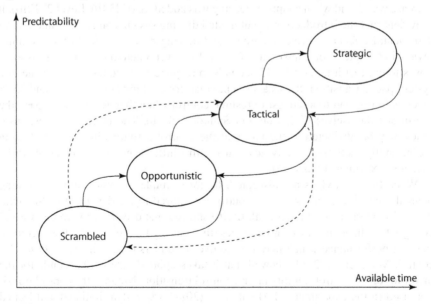

FIGURE 6.2 A finite state model representation of how linear progression through control levels are carried out.

tactical and scrambled, indicating that an operator can move up or down the different control stages more dynamically, depending on time and predictability. As shown in Figure 6.1, the transition to the tactical control mode requires an increase in predictability, indicating that a transition to the tactical level, where decisions can be planned for the near future, requires both sufficient time and predictability to move past the opportunistic or scrambled control level which is highly influenced by the available time. An example in driving where predictability increases the resulting level of control could be losing control of the vehicle on black ice, versus losing control of the vehicle in snowy winter conditions. In the former situation, there is no indication of the reduced friction caused by the black ice, which may result in adverse actions being carried out by the driver, for example by braking or making large steering inputs. In the latter situation, the driver may anticipate the potential for losing control due to the reduced grip in snowy winter conditions and, therefore, for example, steer into the skid, avoid using the brakes and perhaps press down on the throttle to exit the skid safely. These are two situations that have a similar time frame in terms of avoiding going off the road, but the former situation contains enough information for the driver to be able to anticipate the situation and potentially resolve it, moving the driver up the control level from opportunistic to tactical.

Driver assistance systems on Level 1/2 (SAE J3016, 2016) have shown detrimental effects on driver behaviour when drivers were asked to resume control, compared with manual driving. In a meta-analysis, Young and Stanton (2007) observed an increase in brake reaction time of approximately 1 second when a leading vehicle suddenly brakes, requiring driver intervention when using ACC; this is the approximate time it takes a driver to respond to a sudden braking event whilst engaged in fully manual driving (Summala, 2000). An additional 0.3-second increase was found when adaptive steering was added (SAE J3016, Level 2, 2016). It is evident that the introduction of automated driving systems on Level 1 and 2 have detrimental effects on driver readiness and driving performance. In the literature review in Chapter 4 on transitions of control in level 3 automated driving systems, it was found that it takes 1.14–15 seconds to respond to a request to intervene in a system-paced transition. It was also found that drivers took between 1.97 and 25.75 seconds to respond to a request to resume control when they were able to pace the transitions themselves (Eriksson and Stanton, 2017b). The increase in response-times may be attributed to the fact that the control activities involved in driving are normally 'automatic' activities that require little or no conscious effort to be executed (Norman, 1976, p. 70).

When these activities are disrupted, by for example the automation requesting manual control, conscious vehicle control is required to the detriment of 'manual-driving' performance. Russell et al. (2016) showed that drivers who are unaware of changes to vehicle driving characteristics (the steering torque profile) show declined steering performance which may lead to over or undershoot, indicating *scrambled* control. Merat et al. (2014) showed that it takes approximately 40 seconds for the driver to regain control stability after a control transition. Notably, the control transition used in the experiment of Merat et al. (2014) was system-initiated and lacked a Take-Over Request (TOR) that is featured in other recent research into control transitions in automated driving (e.g. Damböck et al., 2012; Eriksson and Stanton,

2017b). The lack of human–machine interface to convey the need to resume control in Merat et al. (2014) may have contributed to *scrambled* control behaviour in the first 40 seconds after resuming control. Similarly, Desmond et al. (1998) found larger heading errors and poorer lateral control in the first 20 seconds after resuming control from automated driving following a failure compared with compensating for a wind gust in manual driving, hinting at *scrambled* control. Moreover, Gold et al. (2013) showed that drivers that were subjected to short lead times (5 vs. 7 seconds) for the TOR elicited shorter response-times, but performance post–take-over was characterised by harsh braking, rapid lane changing and unnecessary full stops, indicating that drivers were at the *scrambled* level of control. Damböck et al. (2012) found that an 8-second lead time for TORs produced driving performance at the same level as manual driving, indicating that drivers experienced a higher *operational* or *tactical* level of control, as according to Hollnagel and Woods (2005) is the expected level of control corresponding to normal human performance. Evidently, there may be a relationship between the available time to resume control from the automated vehicle and the resulting level of control performance when control has been handed back to the driver.

Chapters 4 and 5 argue that 'driver-paced' transitions will be a commonly occurring type of control transition in SAE J3016 (2016) Level 3 and 4 systems, where there is enough foresight when it comes to identifying system limitations (e.g. through the fleet learning feature of Tesla Autopilot version 8. Tesla Motors, 2016), which in turn will increase the lead time between TOR and a transition to manual control (Eriksson and Stanton, 2017b).

The increase in the time between a TOR and a transition to manual control extends the time horizon, which in turn could enable drivers to attain a higher, tactical level of control compared with the system-paced transitions reported in the literature (Russell et al., 2016; Merat et al., 2014; Gold et al., 2013; Damböck et al., 2012; Desmond et al., 1998). Eriksson et al. (Accepted) showed that when the reason for a TOR was highlighted through an augmented reality overlay, drivers exhibited *opportunistic* control by braking to buy time. This was not observed when the augmented reality display showed higher levels of semantics, such as 'arrows' indicating safe paths, indicating a higher level of control in accordance with the maxims in Chapter 2, showing that unambiguous and task-relevant information may improve performance. Hollnagel (1993) emphasise that the 'essence of control is planning' (p.168), which means that a sudden, forced transition to manual control likely has detrimental effects on driving performance unless appropriate support is given (Cranor, 2008; Eriksson and Stanton, 2017a, 2017b; Stanton and Young, 1998).

In light of this, this study aims to explore whether there are any differences in driving after-effects following a transition from automated to manual control in two conditions, one with and one without a secondary task compared with baseline manual driving. This research aims to provide knowledge on whether the resulting level of control (in terms of driving performance) is affected by control transitions when the transition is driver-initiated (Banks and Stanton, 2015, 2016; Lu et al., 2016; Stanton, 2015), driver-paced, and whether the additional time available for self-regulation (Cooper et al., 2009; Eriksson et al., 2014; Kircher et al., 2016; Wandtner et al., 2016) of the transition process has a positive effect on driving performance.

6.1 METHOD

As this chapter is a further analysis of the data used in Chapter 4, a full description of participant demographics, the equipment, procedure and the experimental design is described in full in Sections 4.1.1 through 4.1.4.

6.1.1 DEPENDENT VARIABLES

The following metrics were calculated for each condition and participant for the duration of 20 seconds post-transition to manual control.

- Standard deviation of steering wheel angle (SDSWA, degrees): This metric is related to driver workload. In normal driving conditions, drivers tend to make continuous small steering corrections to adjust their lane position as driving conditions change. When workload increases, these corrections decrease in frequency, resulting in the need for larger steering inputs to correct the lane positioning (Liu et al., 1999). This metric is defined in the Equation 6.1 (c.f. Knappe et al., 2007, p. 2), the equation to calculate standard deviation of the steering wheel angle.

$$\text{SDSWA} = \sqrt{\frac{\sum_{i=1}^{n} \left(SA_i - SA_{\text{avg}} \right)^2}{n}} \tag{6.1}$$

- Mean absolute lateral position (centimetres): This metric describes lane-keeping accuracy and is calculated in the following way:

$$\text{MLP} = \left| \frac{\sum_{i=1}^{n} d_i}{n} \right|$$

where d_i is the distance measured from the centre of the vehicle to the lane centre.

It is argued that these metrics might be a good indicator of the level of control exerted by the driver. For example, higher values of SDSWA, whilst showing large variability in mean absolute lane position, would indicate *scrambled* control; large SDSWA, whilst maintaining low variability in absolute lane position, would indicate *opportunistic* control. Low variability in both metrics would indicate a *tactical* level of control.

In order to determine the after-effects of automated driving with and without a secondary task, before a TOR was issued the mean absolute lateral position and standard deviation of absolute steering wheel angle between the two conditions and a baseline drive was compared. The performance metrics were sampled at 75 Hz and were aggregated into 60 bins (approximately 333.33 … ms of data/bin).

6.2 ANALYSIS

The performance metrics were tested in a time series, assessing the difference between conditions every 333.33 milliseconds for the full 20-second testing period by using t-tests. The tests were corrected using the Bonferroni method, resulting in an alpha of 0.00083, which is represented in the graphs as a red horizontal dashed line in the middle graph of each figure. An uncorrected alpha-level of 0.01 is shown as a horizontal black line in the same graph. The level of significance is represented through the negative base-10 logarithms of P, where large values represent small P-values in a similar fashion to the 'Manhattan' plot (previously used in Gibson, 2010; Petermeijer et al., 2017; Tanikawa et al., 2012). Cohen's D was calculated to assess effect size between conditions in the following way:

$$\text{Cohen's } D = \frac{M_2 - M_1}{SD_{\text{pooled}}}$$

$$SD_{\text{pooled}} = \sqrt{\frac{\left((n_1 - 1) SD_1^2 + (n_2 - 1) SD_2^2\right)}{n_1 + n_2 - 2}} \qquad (6.2)$$

6.3 RESULTS

6.3.1 Mean Absolute Lane Position

When comparing the mean absolute lane position between manual driving and after resuming control from the automation in the passive monitoring condition, the differences in numbers are small (Figure 6.3). This implies that drivers were able to successfully control vehicle lane positioning to the same extent as when driving manually post-resuming control from the automated driving system when able to pace the transition process themselves.

Moreover, similar results can be found when comparing baseline manual driving with driving performance post-resuming control from the automated driving system after being engaged in a secondary task (Figure 6.4), which again suggests that drivers do not suffer from detrimental performance when able to pace the transition process themselves.

Figure 6.5 shows little difference in absolute mean lane position in the two automated driving conditions, from the moment control was resumed to 20 seconds after the moment of resumption of control, suggesting little differences in driving performance post–take-over when drivers are allowed to pace the transition of control themselves. A slight, low magnitude (2–3 cm), non-significant difference in lane position can be seen approximately between 1.5–7 seconds after resuming control.

6.3.2 Standard Deviation of Steering Angle

When comparing the standard deviation of steering angle between the passive monitoring condition after resuming control, with the baseline manual driving condition, there are significant, strong effect differences for the entire time period (Figure 6.6).

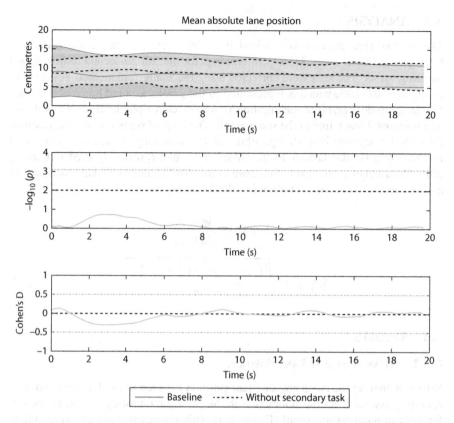

FIGURE 6.3 A time series analysis of the difference in the mean absolute lane position after resuming control from an automated driving system after engaging in passive monitoring compared with manual driving.

This suggests that drivers exerted more effort in maintaining vehicle position through steering corrections than in the manual drive.

Moreover, similar results can be found when comparing the baseline manual driving condition with the secondary task condition (Figure 6.7), again suggesting that drivers had to exert more effort maintaining vehicle position after resuming control than drivers driving manually for a continuous, longer time period.

Lastly, results from the comparison of the standard deviation of steering angle between the passive monitoring, and secondary task condition show a slight, but non-significant increase, in SDSWA in the passive monitoring condition (Figure 6.8), indicating little differences in steering control between the two task conditions when drivers were permitted to pace the transition process themselves.

6.4 DISCUSSION

The aim of this chapter was to assess whether drivers exhibit any of the after-effects in terms of driver control as described by Hollnagel and Woods (2005) through steering behaviour and lane positioning. This was motivated by research suggesting

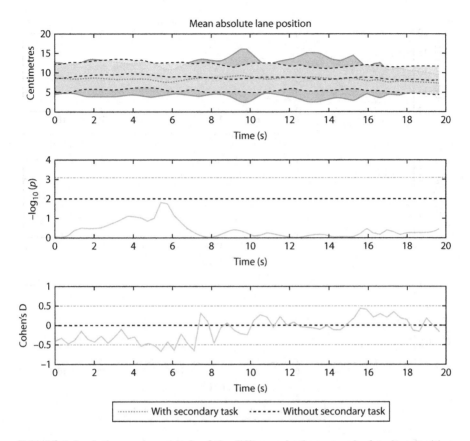

FIGURE 6.4 A time series analysis of the difference in the mean absolute lane position after resuming control from an automated driving system after engaging in a secondary task compared with manual driving.

that drivers adapt their behaviour proactively, depending on the situation and time pressure (Cooper et al., 2009; Eriksson, 2014; Eriksson et al., 2014, Eriksson et al., 2015; Kircher et al., 2014, Kircher et al., 2016; Wandtner et al., 2016). If drivers are allowed to self-pace the transition from automated to manual driving, it was hypothesised that the after-effects and the resulting *scrambled* levels of control, such as harsh braking, sudden lane changes and poor lateral control with large heading errors (e.g. Desmond et al., 1998; Gold et al., 2013, 2016) observed in contemporary research into system-paced transitions of control would be reduced. Indeed, according to Hollnagel and Woods (2005), level of control in a *system* is dependent on the available time, and predictability of a situation, where larger temporal resolution is rewarded with higher, safer levels of control and vice versa. Hollnagel and Woods (2005) also state that the predictability of a situation influences the control mode, and Stanton et al. (2001) state that operators in the *opportunistic* and *tactical* control mode were influenced by the human–machine interface. This implies that the human–machine interface can play a significant role in increasing the predictability of a situation as long as the human–machine interface supplies relevant feedback (as

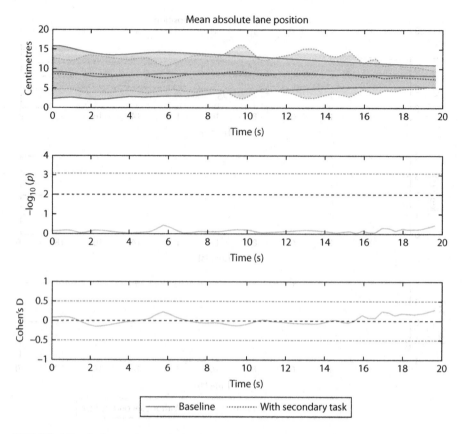

FIGURE 6.5 A time series analysis of the difference in absolute mean lane position after resuming control with a secondary task, and without a secondary task.

suggested in Chapter 2). Hollnagel (1993) emphasises that the 'essence of control is planning' (p. 168), thus implying that forced transitions to manual control likely have a detrimental effect on driving performance.

The results of this study show that when drivers were able to moderate their transition to manual control by means of the time they took to resume control (as reported in Chapter 4), they maintained lateral positioning after the transition to a level comparable to when driving manually in both the passive monitoring (Figure 6.3) and distracted driving condition (Figure 6.4), indicating a higher level of control than what is reported in the literature (c.f. Desmond et al., 1998; Gold et al., 2013, 2016). However, it was also found that drivers exerted more effort maintaining their position in the lane after control was resumed, as shown by the significant increase in standard deviation of steering angle in the two automated conditions compared with the manual driving condition (Figures 6.6 and 6.7 respectively).

The stability of vehicle position, whilst exerting more effort maintaining control performance, indicates that drivers were at the *opportunistic* level of control, approaching the *tactical* level. Research by Russell et al. (2016) found that drivers could not maintain the same level of steering control as during manual driving. An

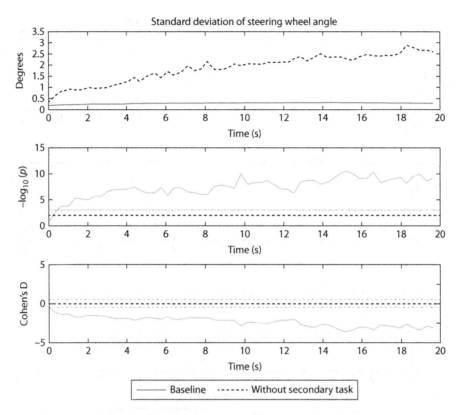

FIGURE 6.6 A time series analysis of the difference in the standard deviation of steering angle after resuming control from an automated driving system after engaging in passive monitoring compared with manual driving.

alternate explanation of the increase in steering control inputs could be that drivers are exploring the vehicle dynamics and whether there has been a change to the dynamics since manual control was ceded. This is somewhat supported by Russell et al. (2016), who hypothesised that drivers would rely on an inaccurate mental model before reaching their previous manual driving performance whilst adapting to the manual driving task to the detriment of normal 'internally automated' performance (Norman, 1976, p. 70). Thus, the increase in standard deviation of steering angle may be attributed to the fact that drivers had to resume a closed loop activity after being removed from the task for some time, and therefore had to show larger steering inputs whilst maintaining lane positioning comparable to manual driving (Russell et al., 2016).

When compared with externally paced transitions in the literature (Russell et al., 2016; Merat et al., 2014; Gold et. al., 2013; Damböck et al., 2012; Desmond et al., 1998), the results obtained in this study show similar effects as reported by Merat et al. (2014), who found that drivers who were asked to resume control in a predictable manner performed better in terms of lateral positioning and steering controllability than drivers who were not expecting a transition to occur, much like in

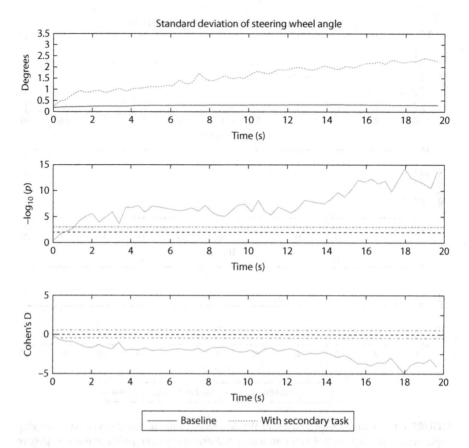

FIGURE 6.7 A time series analysis of the difference in the standard deviation of steering angle after resuming control from an automated driving system after engaging in a secondary task compared with manual driving.

driver-paced versus externally paced transitions of control. The results of this study indicate lesser lane deviation post-TOR for the two automated driving conditions compared with what is reported in Merat et al. (2014). This discrepancy in magnitude between Merat et al. (2014) and the results presented in this chapter could potentially be partially explained by the different bin sizes, as a larger bin size will inherently inflate the standard deviation and may skew the mean due to changes over time. Contrary to Desmond et al. (1998), little effect on lateral positioning was found, indicating that drivers were able to attain an *opportunistic* or *tactical* level of control rather than the seemingly *scrambled* level of control in Desmond et al. (1998). Similarly, Gold et al. (2013) found that drivers exhibited indications of *scrambled* control by performing sudden lane changes or unnecessary use of the hard shoulder, when a short lead time was given. Evidently, there may be a relationship between the available time to resume manual control and the resulting level of control, where higher levels of control are attained when drivers have more time to resume manual control and vice versa.

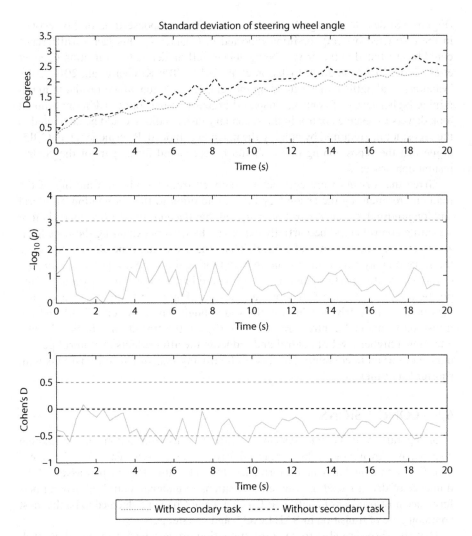

FIGURE 6.8 A time series analysis of the difference in the standard deviation of steering angle after resuming control from an automated driving system after engaging in a secondary task compared with passive monitoring.

This relationship is further strengthened by the results of Gold et al. (2016), who found that drivers performed worse and had more crashes when asked to resume control in road conditions with increased traffic density whilst maintaining a fixed 7-second TORlt. The findings by Gold et al. (2016) serve as an example of a decay to *scrambled* control as traffic density increased. This is in line with Eriksson et al. (2015), who showed that drivers need more time to assess a situation when complexity is high.

When comparing the after-effects of a driver-paced transition on lateral positioning between the two automated driving conditions, no significant differences could be identified in lane positioning (Figure 6.5) and standard deviation of steering angle

(Figure 6.8) despite the significant increase in TOR response-time of 1.6 seconds in the distracted driving condition reported in Chapter 4. This lack of difference could be attributed to the drivers being able to self-moderate the transition (Cooper et al., 2009; Eriksson, 2014; Eriksson et al., 2014, 2015; Kircher et al., 2014, 2016; Wandtner et al., 2016), thus extending the temporal horizon which results in maintaining higher levels of control. Chapter 4 argues that part of the additional time it took drivers to resume control in the secondary task condition could be attributed to the repositioning required by holding the magazine. Indeed, Petermeijer et al. (2015) argue that the repositioning phase should be accounted for as part of the control resumption process.

Given that drivers were completely removed from the visual scanning of the road environment by the secondary task, in addition to the automation of lateral and longitudinal vehicle control, it is argued that the additional 1.6 seconds before resuming control could be partially explained by drivers extending the time horizon to ensure a higher level of control. Thus, the additional time allotted to assess the situation may have been crucial in maintaining the same level of *opportunistic* or *tactical* control performance as was observed in the passive monitoring for lane positioning, and steering behaviour, as shown in Figures 6.5 and 6.8. These findings indicate that drivers who are allowed enough time between a TOR and the transition to manual control are able to self-pace the transition process, thereby attaining a higher level of control and reducing the after-effects on control performance observed in contemporary research utilising system-paced transitions with shorter lead times.

6.5   CONCLUSIONS

Contemporary research into system-paced transitions from automated to manual control has shown indications of reduced driver control levels, leading to detrimental effects on manual driving performance. In light of this, this chapter assessed the influence of driver-paced, non-urgent transitions to manual control on driving performance post-transfer of control, as Chapter 4 and Chapter 5 deemed to be the most common type of transition in SAE Level 3 and 4 systems.

Drivers who were able to pace the transition process back to manual control, irrespective of whether they were engaged in a secondary task or not, exhibited significantly more steering corrections than in manual driving conditions, whilst maintaining comparable positioning in the lane. These findings contradict some of the findings in the contemporary literature, showing that drivers exhibit harsh braking, sudden lane changes and poor lateral control with large heading errors (Desmond et al., 1998; Gold et al., 2013, 2016). This indicates that drivers attained a higher level of control when resuming control in a self-paced setting, compared with an externally paced transfer of control. When the control-transition after-effects were assessed between the passive monitoring and the distracted driving condition, no significant differences in either lateral positioning nor corrective steering behaviour could be found. However, when drivers resumed control in the distracted driving condition, they took an additional 1.6 seconds to do so. This indicates that drivers took more time to complete the transition to extend the time horizon, thus ensuring that a sufficient control level was maintained.

In conclusion, these results indicate that drivers who are given enough time to transfer back to manual control exhibit less of the detrimental effects observed in system-paced conditions (such as those found in Desmond et al., 1998; Gold et al., 2013, 2016). This is promising, as a higher level of vehicle control would reduce the risk of accidents in situations where lead times between a request to resume manual control are required and where manual control is required, such as near the edge of the ODD (SAE J3016, 2016) of the automated driving system (i.e. by a motorway exit, where drivers would be asked to resume control). Such anticipatory behaviour is in line with research by Kircher et al., 2014, who found that users of adaptive cruise control (ACC) anticipated situations where the system would perform poorly and disengaged the system before such a situation could occur, thus strategically pacing their interaction with the system.

This book argues that this type of behaviour is needed from automated vehicles in higher levels of automation (SAE Level 4), where the automation recognises that it is approaching the edge of its ODD and triggers a take-over procedure well in advance to avoid an emergency transition of control. Examples of such behaviour might be to request manual control when approaching a motorway exit, approaching an area where lane markings are faded or even when the system detects that it is approaching an area with poor visibility, such as fog.

It must be noted that the experimental design in this chapter was designed to compress experience and normalise the experience of the transition process to avoid novelty effects, and thus it is not completely representative of the frequency of requests to transition between automated and manual modes that drivers might experience in normal usage of such a system. However, the results support the argument made in Chapter 4 that the 5th–95th percentile range of driver take-over response-time variability must be accounted for in designing the non-urgent take-over process for SAE Level 3 and 4 systems as this would ensure safer transitions to manual vehicle control.

6.6   FUTURE DIRECTIONS

As Chapters 4 through 6 showed that user-paced control transitions are more favourable than system-paced ones in terms of reduction in after-effects and workload, the aim of Chapter 7 is to provide further insights into the control transition process. This will be accomplished by breaking down the process of resuming control into the steps involved in resuming control, as suggested in Petermeijer et al. (2015). Moreover, based on the notion that drivers may be supported through the human–machine interface whilst in opportunistic and tactical control (Hollnagel and Woods, 2005), Chapter 7 aims to determine whether the control transition process can be facilitated through the addition of support and guidance based on the automation taxonomy of Parasuraman et al. (2000) and the Gricean maxims presented in Chapter 2. It is anticipated that this will yield further insights into how drivers resume manual control from an automated vehicle and will generate supporting evidence for the use of the maxims suggested in Chapter 2 in assessing human–machine interfaces.

REFERENCES

Banks, V. A., Stanton, N. A. (2015). Contrasting models of driver behaviour in emergencies using retrospective verbalisations and network analysis. *Ergonomics*, vol 58, no 8, pp 1337–1346.

Banks, V. A., Stanton, N. A. (2016). Keep the driver in control: Automating automobiles of the future. *Applied Ergonomics*, vol 53 Pt B, pp 389–395.

Bye, A., Hollnagel, E., Brendeford, T. S. (1999). Human-machine function allocation: A functional modelling approach. *Reliability Engineering and System Safety*, vol 64, no 2, pp 291–300.

Casner, S. M., Schooler, J. W. (2015). Vigilance impossible: Diligence, distraction, and daydreaming all lead to failures in a practical monitoring task. *Consciousness and Cognition*, vol 35, pp 33–41.

Christoffersen, K., Woods, D. D. (2002). How to make automated systems team players. *Advances in Human Performance and Cognitive Engineering Research*, vol 2, pp 1–12.

Cooper, J. M., Vladisavljevic, I., Medeiros-Ward, N., Martin, P. T., Strayer, D. L. (2009). An investigation of driver distraction near the tipping point of traffic flow stability. *Human Factors*, vol 51, no 2, pp 261–268.

Cranor, L. F. (2008). A framework for reasoning about the human in the loop. Paper presented at the 1st Conference on Usability, Psychology, and Security, San Francisco, CA.

Damböck, D., Bengler, K., Farid, M., Tönert, L. (2012). Übernahmezeiten beim hochautomatisierten Fahren. *Tagung Fahrerassistenz*. München, vol 15, p16.

Dekker, S. W. A., Woods, D. D. (2002). MABA–MABA or abracadabra? Progress on human–automation co-ordination. *Cognition, Technology and Work*, vol 4, no 4, pp 240–244.

Desmond, P. A., Hancock, P. A., Monette, J. L. (1998). Fatigue and automation-induced impairments in simulated driving performance. *Human Performance, User Information, and Highway Design*, no 1628, pp 8–14.

Eriksson, A. (2014). *Driver Behaviour in Highly Automated Driving: An Evaluation of the Effects of Traffic, Time Pressure, Cognitive Performance and Driver Attitudes on Decision-Making Time using a Web Based Testing Platform*. Sweden: Linköping University.

Eriksson, A., Banks, V. A., Stanton, N. A. (2017). Transition to manual: Comparing simulator with on-road control transitions. *Accident Analysis and Prevention*, vol 102, pp 227–234.

Eriksson, A., Lindström, A., Seward, A., Seward, A., Kircher, K. (2014). Can user-paced, menu-free spoken language interfaces improve dual task handling while driving? In M. Kurosu (Ed.), *Human-Computer Interaction. Advanced Interaction Modalities and Techniques* (Vol 8511). Cham, Switzerland: Springer, pp 394–405.

Eriksson, A., Marcos, I. S., Kircher, K., Västfjäll, D., Stanton, N. A. (2015). The development of a method to assess the effects of traffic situation and time pressure on driver information preferences. In D. Harris (Ed.), *Engineering Psychology and Cognitive Ergonomics* (Vol 9174). Cham, Switzerland: Springer International Publishing, pp 3–12.

Eriksson, A., Petermeijer, S. M., Zimmerman, M., De Winter, J. C. F., Bengler, K. J., Stanton, N. A. (Accepted). Rolling out the red (and green) carpet: supporting driver decision making in automation to manual transitions. *IEEE Transactions on Human Machine Systems Subject Corrections*.

Eriksson, A., Stanton, N. A. (2017a). The chatty co-driver: A linguistics approach applying lessons learnt from aviation incidents. *Safety Science*, vol 99, pp 94–101.

Eriksson, A., Stanton, N. A. (2017b). Takeover time in highly automated vehicles: Noncritical transitions to and from manual control. *Human Factors*, vol 59, no 4, pp 689–705.

Gibson, G. (2010). Hints of hidden heritability in GWAS. *Nature Genetics*, vol 42, no 7, pp 558–560.

Gold, C., Damböck, D., Lorenz, L., Bengler, K. (2013). "Take over!" How long does it take to get the driver back into the loop? In *Proceedings of the Human Factors and Ergonomics Society Annual Meeting*, vol 57, no 1, pp 1938–1942.

Gold, C., Korber, M., Lechner, D., Bengler, K. (2016). Taking over control from highly automated vehicles in complex traffic situations: The role of traffic density. *Human Factors*, vol 58, no 4, pp 642–652.

Gordon, M. S., Kozloski, J. R., Kundu, A., Malkin, P. K., Pickover, C. A. (2017). U.S. Patent #9,566,986: Controlling driving modes of self-driving vehicles. http://patft.uspto.gov/netacgi/nph-Parser?Sect1=PTO2&Sect2=HITOFF&p=1&u=%2Fnetahtml%2FPTO%2Fsearch-bool.html&r=1&f=G&l=50&col=AND&d=PTXT&s1=9566986&OS=9566986&RS=9566986.

Hollan, J., Hutchins, E., Kirch, D. (2000). Distributed cognition toward a new foundation for human-computer interaction research. *ACM Transactions on Computer-Human Interaction*, vol 7, no 2, pp 174–196.

Hollnagel, E. (1993). *Human Reliability Analysis: Context and Control*. London: Academic Press.

Hollnagel, E., Woods, D. D. (2005). *Joint Cognitive Systems Foundations of Cognitive Systems Engineering*. Boca Raton, FL: CRC Press.

Hutchins, E. (1995). *Cognition in the Wild*. Cambridge, MA: MIT Press.

Kalra, N., Paddock, S. M. (2016). Driving to safety: How many miles of driving would it take to demonstrate autonomous vehicle reliability? *Transportation Research Part A: Policy and Practice*, vol 94, pp 182–193.

Kircher, K., Eriksson, O., Forsman, Å., Vadeby, A.,, Ahlstrom, C. (2016). Design and analysis of semi-controlled studies. *Transportation Research Part F: Traffic Psychology and Behaviour*, vol 46, no Pt B, pp 404–412.

Kircher, K., Larsson, A., Hultgren, J. A. (2014). Tactical driving behavior with different levels of automation. *IEEE Transactions on Intelligent Transportation Systems*, vol 15, no 1, pp 158–167.

Knappe, G., Keinath, A., Bengler, K., Meinecke, C. (2007). Driving simulator as an evaluation tool - Assessment of the influence of field of view and secondary tasks on lane keeping and steering performance. Paper Presented at the 20th International Technical Conference on the Enhanced Safety of Vehicles (ESV), Lyon, France, June 18–21.

Liu, Y.-C., Schreiner, C. S., Dingus, T. A. (1999). *Development of Human Factors guidelines for Advanced Traveler Information SYstems (ATIS) and Commercial Vehicle Operations (CVO): Human Factors Evaluation of the Effectiveness of Multi-Modality Displays in Advanced Traveler Information Systems* (Monograph).

Lu, Z. J., Happee, R., Cabrall, C. D. D., Kyriakidis, M., de Winter, J. C. F. (2016). Human factors of transitions in automated driving: A general framework and literature survey. *Transportation Research Part F-Traffic Psychology and Behaviour*, vol 43, pp 183–198.

Merat, N., Jamson, A. H., Lai, F. C. H., Daly, M., Carsten, O. M. J. (2014). Transition to manual: Driver behaviour when resuming control from a highly automated vehicle. *Transportation Research Part F-Traffic Psychology and Behaviour*, vol 27, pp 274–282.

Michon, J. A. (1985). A critical view of driver behavior models: What do we know, what should we do? In L. Evans and R. C. Schwing (Eds.), *Human Behavior and Traffic Safety*. New York: Plenum Press, pp 485–524.

Norman, D. A. (1976). *Memory and Attention: An Introduction to Human Information Processing* (2nd Edn). Oxford, UK: Wiley.

Parasuraman, R., Sheridan, T. B., Wickens, C. D. (2000). A model for types and levels of human interaction with automation. *IEEE Transactions on Systems, Man, and Cybernetics - Part A: Systems and Humans*, vol 30, no 3, pp 286–297.

Petermeijer, S. M., Abbink, D. A., de Winter, J. C. (2015). Should drivers be operating within an automation-free bandwidth? Evaluating haptic steering support systems with different levels of authority. *Human Factors: The Journal of the Human Factors and Ergonomics Society*, vol 57, no 1, pp 5–20.

Petermeijer, S. M., Cieler, S., de Winter, J. C. F. (2017). Comparing spatially static and dynamic vibrotactile take-over requests in the driver seat. *Accident Analysis and Prevention*, vol 99, no Pt A, pp 218–227.

Russell, H. E. B., Harbott, L. K., Nisky, I., Pan, S., Okamura, A. M., Gerdes, J. C. (2016). Motor learning affects car-to-driver handover in automated vehicles. *Science Robotics*, vol 1, no 1, pp 1–9.

SAE J3016. (2016). Taxonomy and definitions for terms related to driving automation systems for on-road motor vehicles, J3016_201609: SAE International.

Seppelt, B. D., Victor, T. W. (2016). Potential solutions to human factors challenges in road vehicle automation road vehicle automation. In G. Meyer, S. Beiker (Eds.), *Road Vehicle Automation 3: Lecture Notes in Mobility*. Cham, Switzerland: Springer, pp 131–148.

Stanton, N. A. (2014). Representing distributed cognition in complex systems: How a submarine returns to periscope depth. *Ergonomics*, vol 57, no 3, pp 403–418.

Stanton, N. A. (2015, March). Responses to autonomous vehicles. *Ingenia*, p 9.

Stanton, N. A., Ashleigh, M. J., Roberts, A. D., Xu, F. (2001). Testing hollnagel's contextual control model: Assessing team behavior in a human supervisory control task. *International Journal of Cognitive Ergonomics*, vol 5, no 2, pp 111–123.

Stanton, N. A., Young, M. S. (1998). Vehicle automation and driving performance. *Ergonomics*, vol 41, no 7, pp 1014–1028.

Summala, H. (2000). Brake reaction times and driver behavior analysis. *Transportation Human Factors*, vol 2, no 3, pp 217–226.

Tanikawa, C., Urabe, Y., Matsuo, K., Kubo, M., Takahashi, A., Ito, H. et al. (2012). A genome-wide association study identifies two susceptibility loci for duodenal ulcer in the Japanese population. *Nature Genetics*, vol 44, no 4, pp 430–434, S431–432.

Tesla Motors. (2016). *Upgrading Autopilot: Seeing the World in Radar*. 2017, from https://www.tesla.com/en_GB/blog/upgrading-autopilot-seeing-world-radar?redirect=no

Wandtner, B., Schumacher, M., Schmidt, E. A. (2016). The role of self-regulation in the context of driver distraction: A simulator study. *Traffic Injury Prevention*, vol 17, no 5, pp 472–479.

World Health Organization. (2015). Global Health Observatory (GHO) data: Number of road traffic deaths. from http://www.who.int/gho/road_safety/mortality/number_text/en/

Young, M. S., Stanton, N. A. (2007). Back to the future: Brake reaction times for manual and automated vehicles. *Ergonomics*, vol 50, no 1, pp 46–58.

7 Augmented Reality Guidance for Control Transitions in Automated Driving

Chapter 2 proposed a Chatty Co-Driver approach where the vehicle continuously informs the driver about its state and limitations, as one of the challenges of highly automated driving will be to get a driver who is occupied in a non-driving task back to the driving task when needed (Cranor, 2008). Indeed, a review by De Winter et al. (2014) found that drivers who have been out of the control loop for extended periods of time tend to suffer from degraded situation awareness. Hence, it is of utmost importance to make the driver aware of the functional limitations of the automation before they are reached and an accident can occur (Eriksson and Stanton, 2015, 2016; SAE J3016, 2016; Stanton and Young, 2005). As shown in Chapter 6, drivers can reach higher levels of control by moderating the time needed to transition to manual or automated control. Chapter 6 also noted that drivers can reach higher levels of control if there is high predictability in a situation (and vice versa), and that drivers in the opportunistic and tactical control modes (Hollnagel and Woods, 2005) are dependent on feedback in the human–machine interface (Stanton et al., 2001a). Given the findings of Chapter 6, which indicate a substantial performance increase when drivers were able to pace the transitions on their own accord, it stands to reason that appropriate feedback may have beneficial effects, as indicated by Figure 7.1. Such effects may include bringing the driver into the tactical control level, allowing the driver to make tactical decisions rather than merely focussing on immediate vehicle control (maintaining lateral and longitudinal position on the road). Thus, it is expected that providing feedback to the driver will aid in the forward planning of actions rather than showing improvement in the immediate control needs of the driver.

However, to ensure that the interface assists the driver in reintegrating into the driving control loop, the interface should be designed in accordance with the Gricean maxims proposed in Chapter 2. This will ensure that the interface provides informative, contextually relevant (Eriksson et al., 2015) and unambiguous information that does not overload any sensory channel, taking attention away from the primary task of driving.

Contemporary and near-future highly automated driving systems will require driver intervention, within a 'sufficiently comfortable transition time' (NHTSA, 2013, p. 5) and with 'at least several seconds' after the take-over request (SAE J3016, 2016, p. 6).

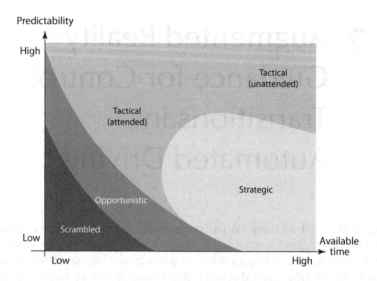

FIGURE 7.1 The relationship between Hollnagel and Woods (2005) control levels and available time and predictability of a situation.

In an attempt to create an understanding of how long drivers need to resume control from an automated vehicle, Chapter 4 reviewed the literature on control transitions and found that drivers take a median of 2.5 seconds, but in some cases even up to 15 seconds to resume control in urgent scenarios (Merat et al., 2014). Chapter 4 also showed that when drivers are requested to resume control without time pressure, they take between 2.1 and 3.5 seconds longer than the reaction times reported in the literature, depending on task engagement (Eriksson and Stanton, 2017b). Moreover, Chapter 4 argued that only considering the 'average driver' is insufficient, because it excludes a large part of the driving population due to the long tail of the reaction time's distribution (see also: Wickens, 2001). According to Petermeijer et al. (2016), reaction times for drivers engaged in non-driving tasks may be faster if additional vibrotactile feedback is provided, compared with visual-only warnings. Scott and Gray (2008) found that tactile warnings lead to a decrease in brake reaction times when compared with no warnings and visual warnings. These effects may be due to the fact that haptic feedback competes less for perceptual resources than visual feedback (Wickens et al., 2015) as driving is primarily a visual task (Sivak, 1996).

Research has indicated that drivers visually assess their environment in preparation for action after receiving a tactile warning (Navarro et al., 2006; Suzuki and Jansson, 2003). The vibrotactile modality is not very effective in conveying complex information and should be primarily used to convey warnings (Gallace and Spence, 2014; Meng and Spence, 2015). Visual displays are, on the contrary, easily understood and can convey complex information (Meng and Spence, 2015), which can be simultaneously processed if is directly linked to the surrounding scene (Foyle et al., 1995). Consequently, the primary use case for visual displays should be to convey context and semantics to the driver, whereas the vibrotactile or auditory could be used to direct attention or provide warnings (Stanton and Edworthy, 1999).

A well-designed human–machine interface should, therefore, pre-cue the driver to resume control using haptic or auditory warnings and provide additional information through the visual or auditory channels to assist the driver in regaining situation awareness.

The issue of how to support the driver by means of feedback or actions is part of an automation framework proposed by Parasuraman et al. (2000). Per this framework, automation can be divided into four stages: information acquisition, information analysis, decision selection and action implementation (in short, *acquisition, analysis, selection* and *implementation*). According to Parasuraman et al. (2000), a technological system can involve different levels of automation within each stage. Note that Parasuraman et al. (2000) based their model on existing models of human information processing, which explains the similarities between the stages of their framework and the stages in the take-over process. An automated vehicle can be categorised in different stages of automation depending on where it lies on the SAE scale (SAE J3016, 2016). A vehicle with Level 2 SAE automation would score highly on information *acquisition* and *analysis* whilst it would score in the lower quarter for decision *selection*, and midway on the scale of action *implementation* (i.e. lack of automatically initiating lane changes and executing such manoeuvres as well as limitations on the force the vehicle can exert on the steering system). On the other hand, a fully automated vehicle (SAE Level 5) would score in the top range on all four scales as the vehicle would be able to alter the route to a destination to accommodate disturbances in traffic such as overtaking slow-moving vehicles. However, if a conditionally automated vehicle (SAE Level 3) reaches its functional limits and presents a take-over request to the driver, this inherently means that the automated system cannot *implement* actions anymore and requires driver intervention to continue to function safely. Despite no longer being able to implement actions, the system could still assist the driver in getting back into the take-over process through a human–machine interface displaying information available from the remaining three stages. A take-over request consisting of a notice in the instrument cluster combined with an auditory signal, as in Gold et al. (2013), would be considered a lower level of *acquisition* (see Figure 7.2; *acquisition – low*). A higher level of *acquisition* (*acquisition – high*) would provide additional information on why it issued the take-over request. Moreover, a user interface that also provides information about the surrounding traffic situation (Stanton et al., 2011) and suggests actions (Zimmermann et al., 2014) would score highly on information *analysis* and decision *selection*, respectively.

The benefits of driver assistance systems have been widely reported in the literature, for example, forward collision warning systems decreased brake reaction times (Banks et al., 2014; Coelingh et al., 2010, 2006) and a haptic gas pedal was found to yield safer car-following behaviour (McIlroy et al., 2016). However, detrimental effects of driver assistance systems have also been reported, primarily in aviation, such as complacency (Parasuraman et al., 1993), skill degradation (Haslbeck and Hoermann, 2016; Wiener and Curry, 1980) and miscommunication between pilot and system (Eriksson and Stanton, 2015, 2017a; Salmon et al., 2016), as shown in Chapter 2.

Another issue that arises with increasing automation across the different levels is automation bias, also known as errors of omission or commission. An error of

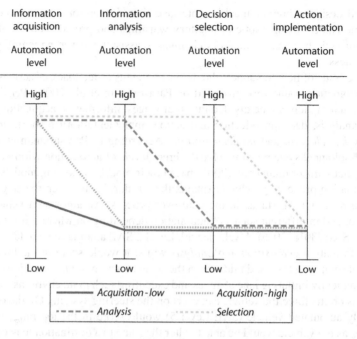

FIGURE 7.2 Qualitative representation of the four levels of information support, namely, acquisition – low (red solid), acquisition – high (blue dotted), analysis (orange long dashes), selection (green short dashes). (Adapted from Parasuraman, R. et al. *IEEE Transactions on Systems, Man, and Cybernetics. Part A, Systems and Human*, 30, 286–297, 2000.)

omission occurs when operators fail to implement an appropriate action because they were not informed by the automated system (Eriksson and Stanton, 2015; Stanton et al., 1997, 2001b, 2011). An error of commission occurs when operators implement a suggested action by the automated system, without considering other indicators, suggesting that the action is incorrect (Skitka et al., 1999); examples of such errors are given in Chapter 2.

Mosier et al. (1996) found that automation bias occurs not only for untrained operators but also for experienced pilots, suggesting that automation bias is a persistent problem for any support system. In highly automated driving, an error of commission could lead to dangerous situations (Stanton and Salmon, 2009) when the system suggests an unsafe action in a take-over scenario. For example, when the system falsely instructs the driver to change lanes whilst the target lane is occupied by other vehicles. Parasuraman et al. (2008) argue that eye-tracking is a useful tool to assess complacency or automation bias. They also mention several studies (Manzey et al., 2006; Metzger and Parasuraman, 2005; Thomas and Wickens, 2004) which show that attention moves away from the primary task when operators are complacent. Similarly, Langois and Soualmi (2016) argued that to give clues about 'what the driver possibly detected and analysed', eye-tracking should be used (p. 1578).

The aim of this experiment was to investigate driver behaviour throughout a take-over scenario. Namely, how driving performance and drivers' gaze behaviour was affected when providing various degrees of augmented reality feedback within the automation levels acquisition, analysis, selection and implementation. It was expected that drivers under higher levels of automation support were more effective (i.e. implemented actions more correctly) and faster (i.e. decreased response-times) in their decision-making process, based on findings by Langlois and Soualmi (2016). Moreover, it was hypothesised that for higher levels of automation, drivers were more prone to automation bias and followed the automation's suggestion more often without verifying the safety of the suggested action (Parasuraman et al., 1993).

7.1 METHOD

7.1.1 Participants

Twenty-five participants (14 male, 11 female) with a mean age of 25.7 years ($SD = 3.9$) and an average driving experience of 8.3 years ($SD = 4.1$) were recruited for participation in this study. The study complied with the American Psychological Association Code of Ethics and had been approved by the University of Southampton Ethics Research and Governance Office (ERGO number: 19930), and all participants provided written informed consent.

7.1.2 Apparatus

A BMW 6-series fixed-base simulator, operating the SILAB (version 4) driving simulator software was used in this experiment (Figure 7.3). It offered a 180° front view and rear projection for every mirror (left, inner and right), generated by six projectors. Road and engine noise were played back, and low-frequency vibrations were provided via a bass shaker in the driver seat. The automation could be toggled by pressing a button (with a diamond-shaped icon) on the steering wheel. Turning the steering wheel more than 2° or pushing the brake pedal disengaged the automation. An icon located between the speedometer and tachometer indicated the automation status (i.e. unavailable, active or inactive).

The participants played 'Angry Birds' as a secondary task during the intervals of highly automated driving. Since it is an interruptible task that does not penalise the player for switching to another task, Angry Birds was deemed suitable. The driver played the game on a Lenovo A7-50 7-inch tablet, which was mounted by the centre console, in front of the radio.

The participants' head and gaze motion were tracked using a three-camera remote system (Smart Eye Pro 6.1; Smart Eye, 2016). Simulation and eye-tracking data were synchronised and logged at 60 Hz. The vehicle environment was modelled in the Smart Eye software to relate gaze and real-world objects: the windshield, rear view mirrors, cluster display and tablet were defined as areas of interest (AOI).

7.1.3 Take-Over Scenario

During each condition, the highly automated vehicle drove in the right lane on a two-lane highway at 110 km/h and approached a slow-moving vehicle (truck, tractor or moped) driving at 58 km/h (see Figure 7.4). When the time to collision (TTC)

FIGURE 7.3   The Technical University Munich Driving Simulator. (Image credit: Fabian Fischer.)

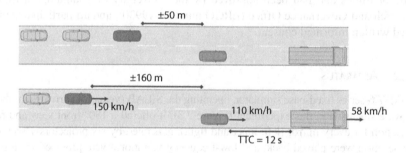

FIGURE 7.4   The two possible take-over scenarios. Top: the group of cars is too close to change lane safely and the driver is expected to brake (i.e. braking scenario). Bottom: the group of cars is far enough away for the driver to overtake the vehicle safely (i.e. steering scenario).

with the obstacle underran 12 seconds, the automation issued a take-over request. Simultaneously, a group of other vehicles, driving at 150 km/h, approached in the left lane. The column was, in the moment of the take-over request, either approximately 160 metres behind (i.e. the first vehicle would pass in approximately 14 seconds), so that the driver could safely change lane (i.e. lane change scenario), or approximately 50 metres behind (i.e. the first vehicle would pass in approximately 4.5 seconds), so that the driver was required to reduce the speed of his or her vehicle (i.e. braking scenario).

7.1.4   HUMAN–MACHINE INTERFACE

To reduce the time needed to respond successfully to a request for manual control, a bi-modal feedback paradigm was utilised. It has previously been shown that such feedback can be used to improve drivers' situational awareness, but this has been shown to have little effect on the immediate control transition process (Lorenz et al., 2014; Narzt et al., 2005). However, research suggests that this type of feedback may aid in decision-making on a tactical rather than operational level (Michon, 1985). The user feedback in this experiment consisted of vibrotactile stimuli in the seat, provided by vibration motors (Figure 7.5). Simultaneously to the vibrotactile warning, an augmented reality display based on Zimmermann et al. (2014) and Lütteken et al. (2016) showed warnings, information or action suggestions for potential courses of action that the driver could take (Figure 7.6). Depending on whether the drivers faced a braking or a lane change scenario, the information analysis and decision selection visuals were redundantly encoded by different colour (red and green; i.e. having a well-established symbolic meaning in a similar fashion to Lorenz et al., 2014), shape (wide or narrow carpet) and direction (left or backwards 'yield' arrows).

This study tested four different types of information support conditions during the take-over scenario with various levels of automation (see Figure 7.2):

1. *Information acquisition – low*: A simple vibrotactile warning indicating that the driver had to resume control. The vibration seat (Figure 7.5) presented a series of three 320 ms pulses (70 ms engaged, 250 ms disengaged) in all 48 motors in the seat to inform the driver he/she needed to resume control. No additional visuals were presented in this condition (Figure 7.6). This vibration was the *baseline* condition.

2. *Information acquisition – high*: At the same moment of the take-over request (i.e. the vibrotactile warning), an augmented sphere highlighted the slowly moving obstacle coming up ahead (Figure 7.6, top right; similar to Zimmermann et al., (2014); Langlois and Soualmi (2016)). The vibration and the sphere overlay were the *sphere* condition.

3. *Information analysis*: In addition to the vibrotactile warning, an augmented reality overlay informed whether there was a gap in the left lane for the

FIGURE 7.5   Illustration of the vibrotactile seat and the location of the 48 vibration motors (i.e. white circles). The motors are arranged in two matrices of 6×4 motors. One in the seat back, the other in the seat bottom.

FIGURE 7.6 The visual interface for the four levels of support. Top left (baseline): no visual support in both scenarios. Top right (sphere): a sphere highlighting the slow-moving vehicle ahead in both scenarios. Middle left (carpet): a carpet in the left lane for the lane change scenario. Middle right (carpet): a barrier covering the lane markings for the braking scenario. Bottom left (arrow): an arrow pointing left for the lane change scenario. Bottom right (arrow): an arrow pointing backwards, for the braking scenario.

driver to change into. A wide, green carpet in the left lane announced the available space in the other lane (like in Zimmermann et al., 2014; Figure 7.6 middle left), whereas a narrow, red road barrier between the lanes emphasised a no-passing zone (inspired by the H-Mode visual feedback; Damböck et al. (2012); Figure 7.6 middle right). The vibration and visual information formed the *carpet* conditions.

4. *Decision selection*: At the same moment of the vibrotactile warning, augmented reality arrows suggested the driver change lane or brake (Figure 7.6 bottom left and right; see Zimmermann et al., 2014). The vibration and the action suggestion displays represented the *arrow* conditions.

7.1.5 EXPERIMENTAL DESIGN

A within-subject, repeated-measures design was used with the conditions counter-balanced for the different user interface solutions. The participants drove four trials with the four support types, namely (1) information acquisition – low, (2) information acquisition – high, (3) information analysis and (4) decision selection. During each trial, the participant experienced six take-over events of which three involved a braking scenario and three involved a lane change scenario (cf. Figure 7.4). Each trial lasted approximately 12 minutes, with a request to resume control due to a braking or lane change scenario every 2 minutes. After each trial, participants were asked to step out of the vehicle to have a break and to fill out two questionnaires.

7.1.6 PROCEDURE

Upon arrival, participants were asked to read an information sheet, containing information regarding the study and the right to at any point abort their trial without any questions asked. After reading the information sheet, the participants were asked to sign an informed consent form. They were also told that they were able to override any system inputs via the steering wheel, throttle or brake pedals. Before commencing the experiment, the participants drove a familiarisation drive to get used to the simulator and the automation they would be using. The drivers were briefed that during the driving scenarios, a number of situations would occur where the car is approaching a slow-moving vehicle ahead. At these moments, the automation would ask the driver to take back manual control of the vehicle. They were also informed that the request to hand back control would be given through three short vibrations in the seat. Moreover, the participants were informed that they would drive four experimental drives testing four different feedback systems, which were described in turn along with images showing how each system would be presented in the environment. At the end of each driving condition, participants were asked to fill out the NASA-RTLX questionnaire (Byers et al., 1989) and the Van Der Laan Technology Acceptance Scale (Van Der Laan et al., 1997). Upon completing an experimental drive, the participants were offered a short break before continuing the study.

7.1.7 DEPENDENT MEASURES

The experiment employed several *objective* measures to capture performance, timing and gaze behaviour, namely:

- *Success rate*: In the lane change scenario, a manoeuvre was considered successful when the driver changed lanes *before* the cars in the adjacent lane passed. In the braking scenario, a manoeuvre was regarded as successful when the participant made a lane change *after* all cars in the adjacent lane had passed.
- *Braking rate*: The percentage of scenarios in which the participants used the brake pedal.
- *Eyes-on-road response-time*: The time it took drivers to move their gaze to the road ahead after a take-over request was issued.

- *Hands-on-wheel response-time*: The time it took drivers to put their hands back on the steering wheel when a take-over request was issued, measured with induction coils in the steering wheel.
- *Steer response-time*: The time it took drivers from the onset of take-over request to the onset of a steering input larger than 2°. Gold et al. (2013) stated that steering wheel angles smaller than 2° are used to stabilise the vehicle and that values larger than 2° can be considered a conscious steering action.
- *Brake response-time*: The time it took drivers from the onset of take-over request to the onset of a depression of the brake pedal larger than 10% as defined in Gold et al. (2013).
- *Lane change time*: The time from the onset of the take-over request to the moment that the participant's vehicle crossed the lane boundary.
- *Gaze behaviour*: The mean and standard deviation of the head heading was used as a measure of the scanning behaviour of the participants. This metric was chosen over the compound vector of the gaze direction and the head heading as the results were comparable, whilst the head heading contained less noise, needing no additional filtering whilst presenting the same result.

In addition, two questionnaires were utilised as *subjective* measures for workload and acceptance:

- The NASA-RTLX was used to evaluate the perceived workload per condition (Byers et al., 1989; Hart and Staveland, 1988).
- A technology acceptance questionnaire (Van Der Laan et al., 1997) was used to measure the usefulness and satisfaction of the different support types.

7.2 ANALYSIS

Due to the non-normal distribution of the control transition data, non-parametric Friedman's tests with a significance (alpha) level of 0.01 were used for the objective measures and 0.05 for the subjective measures. Post-hoc Wilcoxon signed-rank tests (with the alpha level corrected using the Bonferroni correction for multiple comparisons) were used for the objective measures. Effect sizes of the Friedman's test were calculated through Kendall's W test as $N = \chi^2/N(k-1)$, where χ^2 is the Friedman test statistic value, N is the number of participants and k is the number of measurements per participant. For the Wilcoxon signed-rank tests, effect sizes were calculated as $r = |Z|/\sqrt{N}$, where Z is the Z-score obtained from the statistic and N is the number of participants.

For the eye-tracking data, paired t-tests were conducted for each sample at a rate of 60 Hz. The level of significance is represented through the negative base-10 logarithm of p, where large values represent small p-values in a similar fashion to the Manhattan plot, introduced into the Human Factors literature by Petermeijer et al. (2017) and used in Eriksson et al. (accepted), Eriksson and Stanton (2017c),

Gibson (2010), Petermeijer et al. (2017) and Tanikawa et al. (2012). The significance (alpha) level for this measure was set to 0.01. This unconventional use of paired t-tests used for the eye-tracking measures in this study is motivated by the increase in temporal resolution compared with using bin and fewer t-tests. It must be noted that despite conservative corrections of the significance level, the results should only be seen as an indication of an effect, rather than direct evidence. The interpretability of the analysis has been increased through the addition of an effect size measure (Cohen's D) showing the magnitude of the difference between the conditions.

7.3 RESULTS

In 81.3% of the braking scenarios, the participants made a lane change after all the cars in the adjacent lane had passed, whereas in 95.7% of the lane change scenarios, the participants performed a lane change ahead of the cars in the adjacent lane (Table 7.1). A further investigation of visual support conditions indicated that participants changed lane erroneously more often in the baseline and sphere conditions than in the carpet and arrow conditions (Table 7.1).

Similarly, in the lane change scenario, the sphere seemed to yield less successful lane changes and more braking actions compared with the baseline, carpet and arrow.

Figure 7.7 shows the distributions of response-times to the take-over request for eyes road, hands on steering wheel, brake input of more than 10%, steering input of more than 2° and the lane change for the braking scenarios. Figure 7.8 shows the same distributions for the lane change scenarios. The corresponding medians, interquartile ranges (IQRs) and minimum and maximum values for the distributions are provided in Table 7.2.

No significant main effect could be found for the effect of the different visual support systems on first steering input larger than two degrees in both the braking scenario, $\chi^2(3) = 1.42$, $p = 0.702$, $W = 0.02$, and the lane change scenario, $\chi^2(3) = 5.30$, $p = 0.151$, $W = 0.07$. Furthermore, a significant main effect was found for the first

TABLE 7.1

Success Rate and Braking Rate for Steering and Braking Scenarios as a Function of the Support Condition

	Braking Scenario		Steering Scenario	
	Success Rate	Braking Rate	Success Rate	Braking Rate
Baseline	77.3	88.0	96.0	12.0
Sphere	78.7	86.7	86.7	33.3
Carpet	85.3	96.0	100	2.7
Arrow	84.0	96.0	100	6.7
Overall	81.3	91.7	95.7	13.7

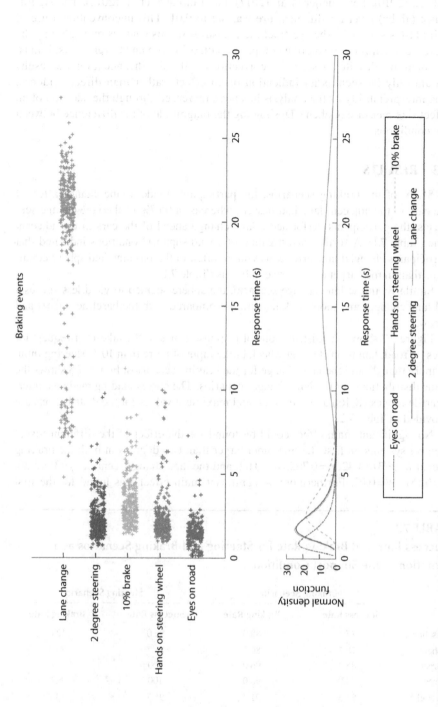

FIGURE 7.7 Braking scenarios. Top: The eyes on road, hands on, brake, steer and lane change response-times during the braking scenarios. Each cross represents the response-time of a single take-over request of a participant. Bottom: The fitted lognormal distributions of the response-times.

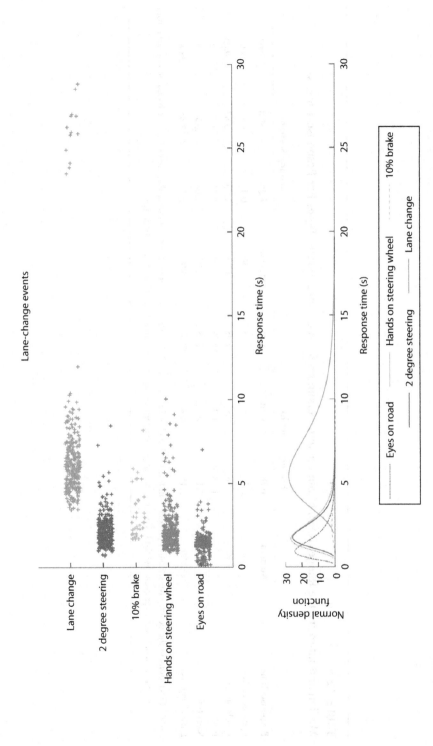

FIGURE 7.8 Steering scenarios. Top: The eyes on road, hands on, brake, steer and lane change response-times during the steering scenarios. Each cross represents the response-time of a single take-over request of a participant. Bottom: The fitted lognormal distributions of the response-times.

TABLE 7.2

Median and Interquartile Range (IQR), Minimum and Maximum for the Two Scenario Types Per Response-Time (s)

Response-Time	Braking Scenarios				Steering Scenarios			
	Median	IQR	Min	Max	Median	IQR	Min	Max
Eyes on road (s)	1.47	0.46	0.13	7.28	1.46	0.42	0.13	7.03
Hands on (s)	1.92	0.88	0.95	9.98	1.95	1.05	0.87	10.05
Brake (s)	2.97	1.83	1.4	11.28	2.68	1.91	1.7	8.2
Steer (s)	2.1	1.18	0.62	9.78	2.02	1.08	0.7	8.5
Lane change (s)	20.89	3.15	4.8	27.4	5.93	2.1	3.45	28.85

Note: Deviations in the descriptive statistics presented in this book and in Eriksson et al. (in press) are not an error, but rather the values here are based on raw data from each individual take-over event, whilst Eriksson et al. (in press) averages take overs across event (lane change/braking).

braking input in the braking scenario, $\chi^2(3) = 14.2$, $p = 0.003$, $W = 0.23$, where braking onset was quicker and harder in the arrow condition (for further info, c.f. Eriksson et al., accepted).

Moreover, a significant main effect of the different information levels, χ^2 $(3, N = 25) = 8.47$, $p = 0.037$, $W = 0.11$, could be found on the time from take-over request to the completion of the lane change in the braking scenario (Figure 7.9, Table 7.3). Post-hoc analysis found a significant difference in lane change execution time between the carpet and the arrow in the braking scenario (Table 7.4).

A significant main effect of the different information levels, $\chi^2(3) = 15.84$, $p = 0.001$, $W = 0.21$ on the time it took to change lane in the lane change scenario was found. Bonferroni-corrected Wilcoxon signed-rank post-hoc tests showed a significantly faster execution of the lane change for the arrow compared with the sphere and baseline as well as for carpet compared with baseline (Table 7.5).

7.3.1 Eye Movements

Figure 7.10 shows the mean head heading angle across participants in the four driving conditions, for the steering and braking scenario. The solid lines show the average between drivers, whereas the shaded area shows the standard deviation measured across participants. Before the take-over request was issued, the heading had an offset towards the right as the participants performed the secondary task that was located on their right. After the take-over request was issued in the lane change scenario, the heading shifted to the left. About 50 metres after the take-over request, the standard deviation across participants dropped, which suggests that the participants focussed on the road ahead. In the braking scenario, it seems that when the arrow was presented, the participants shifted their attention towards the left.

Significantly ($p < 0.01$) larger heading angles to the left were found for the baseline condition compared with the arrow condition, in the braking scenario around 70 metres post–take-over request (Figure 7.11). This effect may be due to the semantic meaning of the arrow efficiently conveying that it was not safe to move into the adjacent lane and therefore reducing the need to check the status of the left lane until the symbol had disappeared. Another potential explanation concerns the location where the arrow was presented. The arrow appeared in the centre of the lane ahead of the driver (Figure 7.6), possibly reducing the perceived need to visually assess the leftmost lane.

7.3.2 Satisfaction and Usefulness Scale

The results of a Friedman's ANOVA showed significant differences in perceived usefulness of the different information support systems $\chi^2(3) = 22.72$, $p < 0.001$, $W = 0.30$. As shown in Figure 7.12 and Table 7.7, sphere scored lowest of all the assessed user interfaces whereas there was a significant increase in satisfaction for arrow and carpet when compared with both baseline and sphere; sphere scored significantly lower than the baseline condition (Table 7.6).

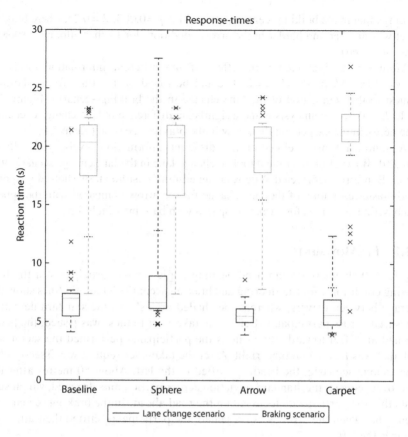

FIGURE 7.9 Adjusted box plot showing the reaction time to a lane change in the braking and steering scenario.

TABLE 7.3
Median and Interquartile Range (IQR), Minimum and Maximum of the Lane Change Time (s) to the Take-Over Request

Reaction	Braking Scenario				Steering Scenario			
Time	Median	IQR	Min	Max	Median	IQR	Min	Max
Baseline (s)	21.03	4.39	4.8	24.15	6.51	2.12	3.95	26.25
Sphere (s)	20.69	3	6.86	24.72	6.38	2.5	4.30	28.85
Carpet (s)	21.33	2.72	7.16	27.4	5.55	2.0	3.45	9.58
Arrow (s)	20.68	3.48	6.05	25.22	5.45	1.14	3.48	8.97

Note: Deviations in the descriptive statistics presented in this book and in Eriksson et al. (accepted) are not an error, but rather the values here are based on raw data from each individual take-over event, whilst Eriksson et al. averages take overs across event (lane change/braking).

TABLE 7.4
Paired Comparisons between the Four Conditions Regarding the Time It Took Drivers to Change Lane after the Take-Over Request in the Braking Scenario

| | Braking Scenario | | | | | | | | |
| | Baseline | | | Sphere | | | Carpet | | |
	Z	p	r	Z	p	r	Z	p	r
Sphere	−0.87	0.382	0.17						
Carpet	2.38	0.017	0.48	1.99	0.046	0.40			
Arrow	0.20	0.840	0.04	1.09	0.276	0.22	−2.92	0.004*	0.58

* Indicates a significant difference at the Bonferroni-corrected Alpha level (0.0083).

TABLE 7.5

Paired Comparisons between the Four Conditions Regarding the Time It Took Drivers to Change Lane after the Take-Over Request in the Steering Scenario

				Steering Scenario							
	Baseline			Sphere				Carpet			
	Z	p	r	Z	p	R		Z	p	r	
Sphere	0.69	0.493	0.14								
Carpet	−2.64	0.008*	0.53	−2.49	0.013	0.50					
Arrow	−3.16	0.002*	0.63	−3.86	<0.001*	0.77		−0.70	0.484	0.14	

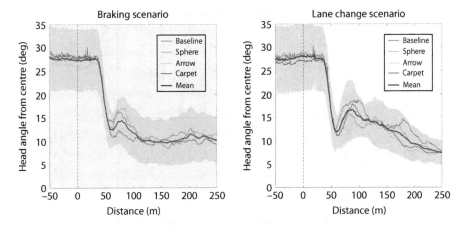

FIGURE 7.10 An overall comparison between the four conditions in the steering and braking scenarios. The black line shows overall mean head heading across conditions.

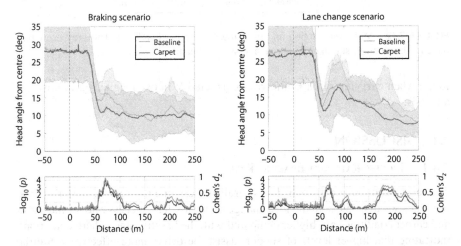

FIGURE 7.11 Top: Mean and standard deviation of (head) heading angle across participants as a function of travel distance for the baseline and arrow conditions. The vertical dashed line indicates the moment of take-over request. Bottom: Paired t-test for the head heading angle between the baseline and arrow conditions.

A Friedman's ANOVA of perceived satisfaction showed significant differences between the different information support systems $\chi^2(3) = 12.30$, $p = 0.006$, $W = 0.16$. Post-hoc Wilcoxon signed-rank analysis using the Bonferroni correction in Table 7.7 showed significant differences between the sphere and the carpet conditions.

7.3.3 OVERALL WORKLOAD

The results of a Friedman's ANOVA showed significant differences in self-reported overall workload (Table 7.8) as measured by the NASA-TLX levels $\chi^2(3) = 10.74$, $p = 0.013$, $W = 0.14$. However, no significant differences could be

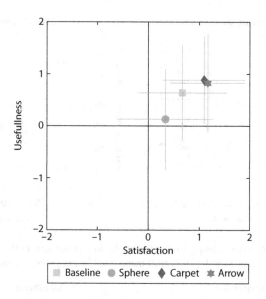

FIGURE 7.12 Mean satisfaction and usefulness score per condition. The grey error bars indicate the mean ±1 standard deviation across participants.

found when carrying out a Bonferroni-corrected Wilcoxon signed-rank post-hoc analysis.

7.4   DISCUSSION

7.4.1   Success Rate and Braking Rate

The results show that drivers had an overall success rate of over 81.3% in the braking scenarios and 95.0% in the lane change scenarios. The success rate for the arrow and carpet conditions are higher compared with the baseline and sphere conditions, indicating that higher levels of support assist the driver more effectively. Similar improvements have been shown by Lorenz et al. (2014), who observed more consistent steering actions when guided by their augmented reality carpets, which is in line with the notion that the human–machine interface can support drivers in gaining a higher level of control (Chapter 6) if it is designed in a way that facilitates the acquisition of Common Ground through utilising the Gricean maxims (Chapter 2). Moreover, Zimmermann et al. (2014) reported higher success rates when employing carpets and arrows.

In the lane change scenario, there is a lower braking rate for arrow and carpet as compared with baseline and sphere, indicating that drivers, supported by higher-level semantic feedback, made their decision to change lane more effectively. The higher percentages of braking in the baseline and sphere conditions in the lane change scenario could potentially be accounted for by the drivers braking due to their uncertainty of actions and to increase their time budget for making a decision. Contrary to expectations, the sphere did little to improve driver decision-making but rather

TABLE 7.6

Paired Comparisons between the Perceived Usefulness and Satisfaction in the Four Conditions

	Baseline			Sphere			Carpet		
	Z	p	R	Z	p	r	Z	p	r
Usefulness									
Sphere	-1.63	0.102	0.33						
Carpet	1.81	0.071	0.36	3.39	<0.001*	0.68			
Arrow	2.99	0.003*	0.60	3.54	<0.001*	0.71	0.61	0.540	0.12
Satisfaction									
Sphere	-1.89	0.059	0.38						
Carpet	0.73	0.464	0.15	3.24	0.001*	0.65			
Arrow	1.09	0.276	0.22	2.40	0.016	0.48	-0.19	0.847	0.04

* Indicates a significant difference at the Bonferroni-corrected Alpha level (0.0083).

TABLE 7.7

Descriptive Statistics for the Van Der Laan Technology Acceptance Scale

	Van Der Laan			
	Baseline M (SD)	Sphere M (SD)	Carpet M (SD)	Arrow M (SD)
Usefulness	0.66 (0.87)	0.33 (0.94)	1.09 (0.8)	1.16 (0.74)
Satisfaction	0.64 (0.92)	0.13 (0.97)	0.88 (0.83)	0.82 (0.95)

TABLE 7.8

Mean and Standard Deviation of the Self-Reported Workload in the Different Experimental Conditions

	Workload			
	Baseline M (SD)	Sphere M (SD)	Carpet M (SD)	Arrow M (SD)
Mental demand	36.8 (22.6)	37.2 (22.6)	31.0 (21.7)	30.2 (22.3)
Physical demand	20.2 (12.8)	21.8 (16.1)	17.8 (14.2)	19.6 (14.9)
Temporal demand	39.4 (20.0)	41.0 (22.3)	31.0 (21.3)	31.4 (22.6)
Performance	30.4 (17.3)	29.2 (17.1)	27.6 (19.8)	27.4 (16.5)
Effort	34.2 (19.8)	28.2 (13.6)	27.6 (18.7)	25.6 (16.5)
Frustration	32.0 (18.1)	34.2 (21.3)	29.0 (21.1)	28.8 (20.4)
Total	**32.2 (14.0)**	**31.9 (13.6)**	**27.3 (15.0)**	**27.2 (15.1)**

degraded it compared with the baseline. It could be that the driver misinterpreted the sphere as a warning of a critical situation ahead, resulting in increased braking. Similarly, Lorenz et al. (2014) found increased braking rates when the road and the slowly moving vehicle ahead were highlighted in red compared with a green carpet indicating a lane change. These findings indicate that merely highlighting an object in the vehicle's path is too ambiguous, and that interfaces need to be designed in a way that is unambiguous and relevant for the situation at hand in accordance with the maxims presented in Chapter 2. Indeed, based on these results, it seems like supporting drivers with the higher levels of semantic feedback could mitigate the instinctive braking response as the display sufficiently informed them of the current situation. These results are also in line with Langlois and Soualmi (2016), who found that drivers would brake more smoothly after a take-over request when they were supported by an augmented reality head-up display.

7.4.2 RESPONSE-TIMES

The non-significant effects between the four conditions regarding the response-times of eyes on road, hands on the steering wheel and steering inputs over 2°, apart from a significantly faster braking response in the arrow condition (for further information c.f. Eriksson et al., accepted), indicate that parts of the initial response are not affected by the information presented to the driver. This is not unexpected as these measures are an indication of the 'repositioning' phase during the take-over scenario, whereas the interface was intended to assist the driver in the 'cognitive processing and action selection'. A possible explanation for the increase and substantial braking in the arrow condition could be that the semantic information provided at the point where the driver is most likely to look after feeling the vibration in the seat was a sufficient cue to initiate braking, whilst, for example, the carpet required a further shift of attention to the left and a confirmatory check before initiating action. Whilst studies by Lorenz et al. (2014) and Langlois and Soualmi (2016) showed no significant differences in initial response-times between two visual interfaces (i.e. augmented red or green carpets) and a control condition, the significant increase in braking input can be explained by the braking arrow being presented in the immediate field of view of the driver as soon as the drivers' attention was directed to the road, whilst providing a clear indication of the necessary action and prompting an appropriate action (i.e. braking in this case).

Consistent with the lower success rates of the baseline and sphere conditions, the arrow yielded faster lane change times than both the sphere and the baseline condition, which exhibited a substantially larger reaction time range in the lane change scenario. A similar trend can be found in the braking scenario, although a decrease of 0.6 seconds is relatively small compared with the approximately 20 seconds it took the drivers to change lane (Table 7.3). The ~1-second decrease between arrow and sphere in the lane change scenario is somewhat more substantial in light of the lane change times of approximately 6 seconds, especially given the substantial difference in range in arrow and sphere (3.48–8.97 seconds and 4.3–28.85 seconds, respectively). These results are in line with research by Damböck et al. (2012), who found that drivers receiving information about the vehicle's trajectory and a lead-vehicle's tracking (in a contact-analogue head-up display) showed faster reactions than those without the support. In this study, these improvements increased further with an increasing level of feedback semantics and colour coding as in the arrow interface concept where shown. Indeed, according to Chapter 2, information should not be more informative than needed (Grice, 1975), but rather aim to convey the intended message with the least amount of effort when it comes to interpreting it (Eriksson and Stanton, 2017a).

Information in the interface should be relevant to the task whilst being presented unambiguously. This may explain the results indicating that the sphere was deemed less usable, producing worse action implementation behaviour than the baseline and the other human–machine interfaces. Moreover, it may also explain why the arrow outperformed the carpet interface, as the human–machine interface elements were placed in areas of the environment where they would be immediately noticed by the driver (i.e. in the lane of the host vehicle), providing semantics highly relevant to the task and required actions without increasing the cognitive effort required to interpret the human–machine interface in accordance with the principles of communication outlined in Chapter 2.

7.4.3 GAZE BEHAVIOUR

The argument that the arrow required less effort to interpret is strengthened by the eye-scanning behaviour in the braking scenario (Figure 7.10, left), indicating that the arrow generated more glances forward compared with the other three levels of feedback. The baseline, sphere and carpet interfaces all show similar gaze behaviour, shifting the attention to the left. The differences in glance behaviour between the carpet and the arrow conditions can be attributed to the lateral offset of the carpet human–machine interface, which would prompt the driver to check whether the barrier still is shown before changing lane, compared with the arrow overlay which is presented directly in the drivers' field of view. Moreover, when comparing the arrow to the baseline condition, the arrow condition reduced the number of glances to the left lane in the braking scenario, whilst generating prompt and strong braking input. It seems that the drivers found the information produced by the human–machine interface (i.e. 'yield arrow') sufficient to determine that they would not change lane and thus found less need to check the adjacent lane. Consequently, it could be argued that the driver showed complacent behaviour in the braking scenario despite the performance increase.

The gaze behaviour in the lane change scenario, however, seemed similar between all conditions (Figure 7.10), indicating that the drivers did check the left lane before changing into it. This indicates that the drivers double-checked the system's suggestion before they implemented an action that could lead to a dangerous situation in the adjacent lane. Contrary to the hypothesis, the participants seemed not to be prone to automation bias. Yet, the participants experienced each of the interaction concepts for a relatively short time-period (approx. 12 minutes), and a longitudinal study should be performed to make any conclusive statements regarding automation bias.

In terms of the perceived usefulness of these interaction concepts, drivers found no difference between the carpet and arrow conditions, further strengthening the case for the use of an appropriate level of semantics in the feedback. Moreover, drivers found both the carpet and the arrow more satisfying than the sphere, which was found to perform worse than the baseline. It was also found that driver satisfaction was higher for the carpet than the sphere, but that carpet and arrow produced similar scores. Zimmermann et al. (2014) also showed that the carpet and arrow were rated higher, especially when they were used together. These results, combined with the small differences in workload between the conditions, indicate that drivers experienced 'equal' assistance from the carpet and arrow, whereas the carpet produced more favourable behaviours in terms of gaze and driver behaviour and reaction times. Moreover, the comparatively worse driving performance in the sphere session combined with the low satisfaction and usefulness scores may indicate that drivers' attention was captured by the user interface elements, but that the information it conveyed was insufficient (with too-high importance), and therefore yielded worse performance as drivers attempted to buy time to interpret the human–machine interface by braking regardless of whether a lane change was necessary or not, leading to lower satisfaction and usefulness scores.

7.5 CONCLUSION

The study described in this chapter assessed four types of human–machine interfaces with increasing levels of semantic meaning, classified along the stages of automation suggested by Parasuraman (2000). It was hypothesised that drivers would benefit from the data acquired by vehicle sensors by automating *information acquisition* and *information analysis* to aid *decision selection* before and during transitions to manual vehicle control following highly automated driving. That the user interfaces assessed in this chapter would help with the initial driver response to the take-over request (i.e. the shift of attention and the repositioning phase of the hands) was neither expected nor found (as the initial reaction times failed to show significance).

Significant improvements appeared however when assessing the effect of the user interface in the cognitive processing and action selection phase. Drivers had to assess whether to brake or to change lane after the request to resume control in this phase. Merely highlighting an obstacle did little to improve driver behaviour, but rather it increased the amount of unnecessary braking, as drivers slowed down seemingly to gain more time to assess the situation when the human–machine interface was insufficient, in accordance with the Contextual Control Model (COCOM), and as described in Chapter 6. This behaviour was not as prominent in the baseline condition.

Moreover, when drivers experienced the carpet or arrow interaction concepts, a substantial improvement in correct and timely action implementation was observed (between 3.45–9.58 seconds in the carpet and 3.48–8.97 seconds in the arrow condition compared with 3.95–26.25 seconds in the baseline condition and 4.3–28.85 seconds in the sphere condition). This could be a result of the support of the interface in the cognitive processing and action selection phases. Out of the two higher-level, semantic interfaces, the carpet and arrow outperformed the sphere and baseline concepts in terms of response-time where the arrow proved to be marginally faster.

However, the gaze behaviour in the braking scenario indicated complacent behaviour of the driver (i.e. not double-checking if the adjacent lane was indeed blocked). In light of these results, it seems that the relationship – when it comes to improving driver performance using increasing levels of information support – is not as straightforward as hypothesised. That is, the effect of a system on driver performance is not only dependent on the level of information that it provides but is dependent on many factors like the salience and location of the provided support, as previously argued in Chapter 2. Moreover, providing the driver with low-level support might be detrimental to the response of drivers, since they have more information to interpret than they would in a baseline scenario (Eriksson and Stanton, 2017a). Interface designers should take these findings into account since it is not easy to predict whether an interface will have the intended beneficial effect or whether it might be detrimental to performance.

It is, therefore, in accordance with Chapter 2, recommended to maintain salient feedback in the relevant location, that is, arrows in the host-lane where the initial attention of the driver will be directed to in a take-over event. Additionally, feedback conveying such high-level, semantic information, that is, the arrows presented in the driver's immediate field of view, should be implemented with care, as some support for automation bias being present was observed.

Lastly, drivers rated the overall workload experienced in the take-over scenarios lower when supported by the high semantic interfaces. This further leads to the notion that a utilisation of vehicle sensor data may be useful to drivers on a tactical level through aiding decision-making – despite lacking the option to drive automated. In this study, it was shown that driver decision-making can be aided. Thus, there are indications that the human–machine interface is able to increase driver control from opportunistic to tactical as long as the interface provides enough information in the relevant locations (i.e. carpet on the relevant area of the roadway or arrows indicating potential actions to take to resolve the situation), allowing the driver to plan ahead. This results in faster and more appropriate actions with reduced workload for the driver in a scenario where a slowly moving vehicle is blocking the lane and the driver may choose between changing lane or reducing vehicle velocity. It has previously been shown that drivers tend to change their information needs as situations change (Eriksson et al., 2015; Walker et al., 2015). Therefore, future research on how to design human–machine interaction to improve feedback and aid decision-making should be carried out in a larger variety of situations.

REFERENCES

Banks, V. A., Stanton, N. A., Harvey, C. (2014). What the drivers do and do not tell you: Using verbal protocol analysis to investigate driver behaviour in emergency situations. *Ergonomics*, vol 57, no 3, pp 332–342.

Byers, J. C., Bittner, A., Hill, S. (1989). Traditional and raw task load index (TLX) correlations: Are paired comparisons necessary. In A. Mital (Ed.), *Advances in Industrial Ergonomics and Safety* (vol 1). London: Taylor & Francis, pp 481–485.

Coelingh, E., Eidehall, A., Bengtsson, M. (2010). Collision warning with full auto brake and pedestrian detection - A practical example of automatic emergency braking. In *13th International IEEE Conference on Intelligent Transportation Systems (ITSC)*, pp 155–160.

Coelingh, E., Lind, H., Birk, W., Distner, M., Wetterberg, D. (2006). Collision warning with auto brake. Paper presented at the FISITA 2006 World Automotive Congress, October 22–27.

Cranor, L. F. (2008). A framework for reasoning about the human in the loop. In *Proceedings of the 1st Conference on Usability, Psychology, and Security*, San Francisco, CA.

Damböck, D., Weißgerber, T., Kienle, M., Bengler, K. (2012). Evaluation of a contact analog head-up display for highly automated driving. Paper presented at the 4th International Conference on Applied Human Factors and Ergonomics, San Francisco, CA.

De Winter, J. C., Happee, R., Martens, M. H., Stanton, N. A. (2014). Effects of adaptive cruise control and highly automated driving on workload and situation awareness: A review of the empirical evidence. *Transportation Research Part F: Traffic Psychology and Behaviour*, vol 27, pp 196–217.

Eriksson, A., Marcos, I. S., Kircher, K., Västfjäll, D., Stanton, N. A. (2015). The development of a method to assess the effects of traffic situation and time pressure on driver information preferences. In D. Harris (Ed.), *Engineering Psychology and Cognitive Ergonomics* (vol. 9174). Cham, Switzerland: Springer, pp 3–12.

Eriksson, A., Petermeijer, S., Zimmerman, M., De Winter, J. C. F., Bengler, K., Stanton, N. A. (in press). Rolling out the red (and green) carpet: Supporting driver decision making in automation to manual transitions. *IEEE Transactions on Human Machine Systems*.

Eriksson, A., and Stanton, N. A. (2015). When communication breaks down or what was that? – The importance of communication for successful coordination in complex systems. Paper presented at the 6th International Conference on Applied Human Factors and Ergonomics, Las Vegas, NV.

Eriksson, A., Stanton, N. A. (2016). The chatty co-driver: A linguistics approach to human-automation-interaction. Paper presented at the Ergonomics and Human Factors (IEHF) Conference, Daventry, UK.

Eriksson, A., Stanton, N. A. (2017a). The chatty co-driver: A linguistics approach applying lessons learnt from aviation incidents. *Safety Science*, vol 99, pp 94–101.

Eriksson, A., Stanton, N. A. (2017b). Takeover time in highly automated vehicles: Noncritical transitions to and from manual control. *Human Factors*, vol 59, no 4, pp 689–705.

Eriksson, A., Stanton, N. A. (2017c). Driving performance after self-regulated control transitions in highly automated vehicles. *Human Factors,* vol 59, no 8, pp 1233–1248.

Foyle, D. C., McCann, R. S., Shelden, S. G. (1995). Attentional issues with superimposed symbology: Formats for scene-linked displays. In *Proceedings of the Eighth International Symposium on Aviation Psychology*, Columbus, OH.

Gallace, A., Spence, C. (2014). *In Touch with the Future: The Sense of Touch from Cognitive Neuroscience to Virtual Reality*. Oxford: Oxford University Press.

Gibson, G. (2010). Hints of hidden heritability in GWAS. *Nature Genetics*, vol 42, no 7, pp 558–560.

Gold, C., Damböck, D., Lorenz, L., Bengler, K. (2013). "Take over!" How long does it take to get the driver back into the loop? In *Proceedings of the Human Factors and Ergonomics Society Annual Meeting*, vol 57, no 1, pp 1938–1942.

Grice, H. P. (1975). Logic and conversation. In P. Cole and J. L. Morgan (Eds.), *Speech Acts.* New York, NY: Academic Press, pp. 41–58.

Hart, S. G. and Staveland, L. E. (1988). Development of NASA-TLX (Task Load Index): Results of empirical and theoretical research. *Advances in Psychology*, vol 52, pp 139–183.

Haslbeck, A., Hoermann, H. J. (2016). Flying the needles: Flight deck automation erodes fine-motor flying skills among airline pilots. *Human Factors*, vol 58, no 4, pp 533–545.

Hollnagel, E., Woods, D. D. (2005). *Joint Cognitive Systems Foundations of Cognitive Systems Engineering.* Boca Raton, FL: CRC Press.

Langlois, S.,Soualmi, B. (2016). Augmented reality versus classical HUD to take over from automated driving: An aid to smooth reactions and to anticipate maneuvers. In *IEEE 19th International Conference on Intelligent Transportation Systems (ITSC)*. Rio de Janeiro, Brazil: IEEE, pp 1571–1578.

Lorenz, L., Kerschbaum, P., Schumann, J. (2014). Designing take over scenarios for automated driving how does augmented reality support the driver to get back into the loop? In *Proceedings of the Human Factors and Ergonomics Society Annual Meeting*, vol 58, no 1, pp 1681–1685.

Lütteken, N., Zimmermann, M., Bengler, K. (2016). Using gamification to motivate human cooperation in a lane-change scenario. In *IEEE 19th International Conference on Intelligent Transportation Systems (ITSC)*. Rio de Janeiro, Brazil: IEEE.

Manzey, D., Bahner, J. E., Hueper, A.-K. (2006). Misuse of automated aids in process control: Complacency, automation bias, and possible training interventions. In *Proceedings of the Human Factors and Ergonomics Society 50th Annual Meeting,* Santa Monica, CA: Human Factors.

McIlroy, R. C., Stanton, N. A., Godwin, L., Wood, A. P. (2016). Encouraging eco-driving with visual, auditory, and vibrotactile stimuli. *IEEE Transactions on Human-Machine Systems*, vol 47, no 5, pp 661–672.

Meng, F., Spence, C. (2015). Tactile warning signals for in-vehicle systems. *Accident Analysis and Prevention*, vol 75, pp 333–346.

Merat, N., Jamson, A. H., Lai, F. C. H., Daly, M., Carsten, O. M. J. (2014). Transition to manual: Driver behaviour when resuming control from a highly automated vehicle. *Transportation Research Part F – Traffic Psychology and Behaviour*, vol 27, pp 274–282.

Metzger, U., Parasuraman, R. (2005). Automation in future air traffic management: Effects of decision aid reliability on controller performance and mental workload. *Human Factors*, vol 47, no 1, pp 35–49.

Michon, J. A. (1985). A critical view of driver behavior models: What do we know, what should we do? In L. Evans and R. C. Schwing (Eds.), *Human Behavior and Traffic Safety*. New York, NY: Plenum Press, pp. 485–524.

Narzt, W., Pomberger, G., Ferscha, A., Kolb, D., Müller, R., Wieghardt, J., et al. (2005). Augmented reality navigation systems. *Universal Access in the Information Society*, vol 4, no 3, pp 177–187.

Navarro, J., Mars, F., Hoc, J.-M., Boisliveau, R., Vienne, F. (2006). Evaluation of human-machine cooperation applied to lateral control in car driving. In *Proceedings of the 16th World Congress of the International Ergonomics Society*, Maastricht, the Netherlands, pp 4957–4962.

NHTSA (National Highway Traffic Safety Administration). (2013). *Preliminary Statement of Policy Concerning Automated Vehicles*. NHTSA.

Parasuraman, R., Molloy, R., Singh, L. I. (1993). Performance consequences of automation-induced 'complacency'. *The International Journal of Aviation Psychology*, vol 3, no 1, pp 1–23.

Parasuraman, R., Sheridan, T. B., Wickens, C. D. (2000). A model for types and levels of human interaction with automation. *IEEE Transactions on Systems, Man, and Cybernetics. Part A, Systems and Humans*, vol 30, no 3, pp 286–297.

Parasuraman, R., Sheridan, T., Wickens, C. D. (2008). Situation awareness, mental workload, and trust in automation: Viable, emperically supported cognitive engineering constructs. *Journal of Cognitive Engineering and Decision Making*, vol 2, no 2, pp 140–160.

Petermeijer, S. M., Cieler, S., de Winter, J. C. F. (2017). Comparing spatially static and dynamic vibrotactile take-over requests in the driver seat. *Accident Analysis and Prevention*, vol 99, Pt A, pp 218–227.

Petermeijer, S. M., de Winter, J. C. F., Bengler, K. J. (2016). Vibrotactile displays: A survey with a view on highly automated driving. *IEEE Transactions on Intelligent Transportation Systems*, vol 17, no 4, pp 897–907.

SAE J3016. (2016). Taxonomy and definitions for terms related to driving automation systems for on-road motor vehicles, *J3016_201609*. SAE International.

Salmon, P. M., Walker, G. H., Stanton, N. A. (2016). Pilot error versus sociotechnical systems failure: a distributed situation awareness analysis of Air France 447. *Theoretical Issues in Ergonomics Science*, vol 17, no 1, pp 64–79.

Scott, J. J., Gray, R. (2008). A comparison of tactile, visual, and auditory warnings for rear-end collision prevention in simulated driving. *Human Factors*, vol 50, no 2, pp 264–275.

Sivak, M. (1996). The information that drivers use: Is it indeed 90% visual? *Perception*, vol 25, no 9, pp 1081–1089.

Skitka, L. J., Mosier, K. L., Burdick, M. (1999). Does automation bias decision-making? *International Journal of Human-Computer Studies*, vol 51, no 5, pp 991–1006.

Stanton, N. A., Ashleigh, M. J., Roberts, A. D., Xu, F. (2001a). Testing hollnagel's contextual control model: Assessing team behavior in a human supervisory control task. *International Journal of Cognitive Ergonomics*, vol 5, no 2, pp 111–123.

Stanton, N. A., Young, M. S., Walker, G. H., Turner, H., Randle, S. (2001b). Automating the driver's control tasks. *International Journal of Cognitive Ergonomics*, vol 5, no 3, pp 221–236.

Stanton, N. A., Dunoyer, A., Leatherland, A. (2011). Detection of new in-path targets by drivers using Stop & Go Adaptive Cruise Control. *Applied Ergonomics*, vol 42, no 4, pp 592–601.

Stanton, N. A., Edworthy, J. (1999). *Human Factors in Auditory Warnings.* Aldershot, UK: Ashgate.

Stanton, N. A., Salmon, P. M. (2009). Human error taxonomies applied to driving: A generic driver error taxonomy and its implications for intelligent transport systems. *Safety Science*, vol 47, no 2, pp 227–237.

Stanton, N. A., Young, M. S. (2005). Driver behaviour with adaptive cruise control. *Ergonomics*, vol 48, no 10, pp 1294–1313.

Stanton, N. A., Young, M. S., McCaulder, B. (1997). Drive-by-wire: The case of mental workload and the ability of the driver to reclaim control. *Safety Science*, vol 27, no 2–3, pp 149–159.

Suzuki, K., Jansson, H. (2003). An analysis of driver's steering behaviour during auditory or haptic warnings for the designing of lane departure warning system. *JSAE Review*, vol 24, no 1, pp 65–70.

Tanikawa, C., Urabe, Y., Matsuo, K., Kubo, M., Takahashi, A., Ito, H. et al. (2012). A genome-wide association study identifies two susceptibility loci for duodenal ulcer in the Japanese population. *Nature Genetics*, vol 44, no 4, pp 430–434, S1–2.

Thomas, L. C., Wickens, C. D. (2004). Eye-tracking and individual differences in off-normal event detection when flying with a synthetic vision system display. In *Proceedings of the Human Factors and Ergonomics Society 48th Annual Meeting,* Santa Monica, CA: Human Factors and Ergonomics Society.

Van Der Laan, J. D., Heino, A., De Waard, D. (1997). A simple procedure for the assessment of acceptance of advanced transport telematics. *Transportation Research Part C: Emerging Technologies*, vol 5, no 1, pp 1–10.

Walker, G. H., Stanton, N. A., Salmon, P. M. (2015). *Human Factors in Automotive Engineering and Technology.* Farnham, UK: Ashgate.

Wickens, C. D. (2001). Attention to safety and the psychology of surprise. In *Proceedings of the 2001 Symposium on Aviation Psychology*, Columbus, OH: The Ohio State University.

Wickens, C. D., Hollands, J. G., Banbury, S., Parasuraman, R. (2015). *Engineering Psychology and Human Performance.* Abingdon, UK: Psychology Press.

Wiener, E. L., Curry, R. E. (1980). Flight-deck automation: Promises and problems. *Ergonomics*, vol 23, no 10, pp 995–1011.

Zimmermann, M., Bauer, S., Lütteken, N., Rothkirch, I. M., Bengler, K. J. (2014). Acting together by mutual control: Evaluation of a multimodal interaction concept for cooperative driving. In *International Conference on Collaboration Technologies and Systems (CTS)*, Minneapolis, MN: IEEE.

8 Conclusions and Future Work

8.1 INTRODUCTION

The aim of the research disseminated in this book was to study driver interaction with highly automated vehicles, through a multidisciplinary approach applying lessons learnt in other domains, using driving simulations and on-road experimentation. The main findings are discussed in the sections that follow, along with a discussion regarding the contributions and implications, as well as limitations, of the research approach. Lastly, potential avenues for future enquiry are discussed.

8.2 SUMMARY OF FINDINGS

The research presented in this book was structured around five key objectives, the findings of which are summarised here.

1. Propose a framework for reasoning about human–automation interaction, based on linguistic theories on human–human communication.

   Chapter 2 introduced a novel way of reasoning about humans interacting with automation that is dubbed 'a Chatty Co-Driver paradigm', and is based on the linguistic principles of human-to-human interaction. It was proposed that the maxims of successful conversation proposed by Paul Grice (1975) be used to assess human–automation interaction. Ensuring efficient communication between the driver and the vehicle is deemed to be of utmost importance as the driver is further removed from the driving loop in contemporary and future automated vehicles. As the driver is removed from the driving loop, whilst expected to intervene when the vehicle deems it can no longer cope with a situation, it falls on the system designer to design feedback that can efficiently guide the driver back into the control loop. It can be argued that whilst the vehicle requests manual driving, it still has access to a plethora of sensors that are continuously acquiring information about the vehicle surroundings; such information could be conveyed to the driver to facilitate the re-integration into the driving loop. This notion was further tested in Chapter 7, where different types of feedback on the four levels of automation (Parasuraman, 2000) were used to guide the quality of information available to the vehicle.

   This feedback was then shown to the driver in different human–machine interface configurations that fulfil different aspects of the Gricean maxims of human–human communication. Chapter 2 concludes that to reduce the gulf of evaluation in human–automation systems, Human Factors engineers must ensure that relevant information is successfully communicated to drivers in an appropriate fashion, as driving, much like aviation, is a domain where communication breakdowns are a serious threat to safety.

2. Design Open Source software algorithms that will allow automated driving by a generic driving simulator, in a way that enables dynamic driver interaction with the system, and will also allow us to assess whether the behaviours observed in the simulator can be found in on-road driving conditions, thus lending validity to the simulator as a tool for automated driving research.

Chapter 3 introduced the software algorithms developed to enable the Southampton University Driving Simulator to conduct research into highly automated vehicles. This software has been a core component in conducting the research presented in this book and was used in Chapters 4 and 6 to assess driver interaction with automated driving systems. The findings of Chapter 4 were later compared with on-road driving conditions in Chapter 5, which showed a high level of correlation between the behaviours observed on the road and those observed in the simulator, indicating a successful reproduction of the automated driving experience by the algorithms. These algorithms have now been released as an Open Source software toolbox, thus enabling the STISIM community to utilise the software to research automated driving Human Factors on a flexible and validated platform which allows direct comparison between studies.

3. Validate the Southampton University Driving Simulator and the algorithms in Chapter 3 against driver behaviour during real-world driving conditions.

This book provided behavioural validation (relative validity) of the Southampton University Driving Simulator for the area of automated driving research. Chapter 4 assessed the take-over times in a simulated highly automated vehicle, and Chapter 5 assessed the correspondence of the findings in Chapter 4 through a correlation analysis of an on-road trial using a sample matched on age and driving experience.

The chapter found strong correlations for the transition times to, and from, automated vehicle control ($r > 0.96$ and $r > 0.97$ respectively). These findings are of great value to ongoing research as well as future research ventures. They also support the findings in Chapters 4 through 6 as they show that the simulator produces results corresponding to those of on-road conditions. As stated by Rolfe et al. (1970, p. 761), 'The value of a simulator depends on its ability to elicit from the operator the same sort of response that he would make in the real situation'.

4. Provide design guidance for vehicle manufacturers and guidelines for policymakers on the transition process between automated and manual driving based on experimental evidence.

Design guidelines for driver interaction have been proposed based on the theoretical findings in Chapter 2, through the application of the

Gricean maxims of human–human interaction to the interaction between driver and automated driving systems based on case studies in aviation, where automation is prevalent. Further recommendations were then generated based on the experimental findings of Chapter 7, showing results in line with the theoretical predictions in Chapter 2 (i.e. contextually relevant feedback produced better driver decisions than other feedback). Furthermore, design recommendations for the take-over process were made in Chapters 4 through 6, advising vehicle manufacturers and policy-makers to accommodate the full range of driver performance in a similar fashion to what is being done in anthropometrics, that is, accommodating the 5th percentile female to the 95th percentile male, rather than focus-sing on the average driver response-times. Further recommendations were made for the handover process advising vehicle manufacturers (in accordance with the SAE guidelines SAE J3016, 2016) to provide drivers with plenty of time before a transition of control is made if possible, as this seems to have a positive effect on driving performance in the time after control has been transferred back to the driver. Lastly, a first-of-a-kind finding regarding the time it takes drivers to respond to a request to engage automated driving was disseminated in the academic litera-ture (Eriksson and Stanton, 2017b); it was later validated in on-road trials (Eriksson et al., 2017).

8.3 IMPLICATIONS OF RESEARCH

8.3.1 THEORETICAL

This book offers a novel application of linguistics to the Human Factors domain, more specifically in how humans interact with automated agents such as in auto-mated cars. It adds to the sparse body of literature regarding communications theory used with automated vehicles and offers an alternate approach in assessing feedback in human–machine systems by assessing the interaction in the same way as for human–human interaction. Parts of these findings were then applied in the driving simulator to assess different human–machine interfaces with results supporting the use of linguistic theory in the assessment of driver-automation interaction.

Moreover, this book provides a novel approach for looking at the control tran-sition process (Chapters 4 and 6) and how long drivers need to resume control from an automated vehicle; it offers a viewpoint different from that of the body of contemporary literature on the topic. It was suggested that the full range of driver variability should be accommodated (i.e. 5th percentile driver to 95th percentile driver, as is standard in ergonomics and anthropometrics), rather than focussing on the average driver, as currently is done in contemporary research. Throughout this book, we have seen that allowing driver-paced transitions of control can benefit the control transition process by maintaining workload on an optimal level whilst having little detrimental effect on post–take-over driving performance.

8.3.2 METHODOLOGICAL

This book has contributed to three methodological areas, first in terms of developing the first-ever implementation of automated driving made publicly available for the STISIM driving simulator that allows flexible interaction with the system. This implementation allows research into more naturalistic behaviours where the interaction between driver and automation may be studied. An example of such dynamic interaction is how drivers engage and disengage the system based on surrounding environmental factors such as traffic, roadworks or entering urban environments. This type of study relates to the second methodological contribution of this book to the area of design and analysis of semi-controlled studies (Kircher et al., 2016), which is in its infancy in the area of automotive research.

This book adds to the relatively small body of literature employing semi-controlled study designs, whilst highlighting its strengths (i.e. by showing that drivers exhibit more variance in response-time when they are self-regulating the process of resuming control from an automated vehicle, which differs from the literature utilising fixed time budgets). Lastly, this book provided a validation of the Southampton University Driving Simulator. The results in Chapter 5 show high correlations for transitions from automated to manual control, and vice versa ($r=0.97$ and $r=0.96$ respectively). These results showed relative validity for the task of transferring control, and thus the research into automated driving presented in this book, and future studies carried out in the Southampton University Driving Simulator.

This is an important finding as there is little research being conducted (and published) on automated driving in on-road conditions. As stated in Chapter 5, most contemporary research is carried out in the simulator, on sub-systems of automation or on closed test tracks. The finding of strong correlations between simulated driving and driving in real traffic conditions on public roads lends validity to the research into control transition in automated driving already disseminated in the contemporary literature.

8.3.3 PRACTICAL

It is important to consider the practical implications of the research presented in this book. One of the key areas for consideration is the application of the linguistic theory of human–human interaction presented in Chapter 2 in generic automotive interfaces as well as in interfaces to convey information related to the automated vehicle status. Moreover, there is no reason why this theoretical framework cannot be applied to generic human–agent systems, such as in aviation automation, in the design of human–machine interfaces in medical devices, process control, maritime environments or robotics (Thellman, 2016). Another area for consideration is the notion of reintegrating the accommodation of inter- and intra-human variability back into Human Factors, and not limiting it to the physical Ergonomics/Human Factors domain as is the current practice. This is of particular importance when designing safety-critical systems, where operators act as crucial fallback components. It is anticipated that these findings will help facilitate the interaction between humans and other agents by accommodating human variability as well as designing the interaction in a way that mimics that of human–human communication.

8.4 FUTURE WORK

Several opportunities for further academic enquiry have been raised as a result of the research findings presented in this book. These opportunities are summarised here.

8.4.1 AUTOMATION AND THE EFFECT OF EXTERNAL FACTORS ON THE CONTROL TRANSITION PROCESS

Whilst there is some research exploring the effects of external factors on the control-transition process (Gold et al., 2016; Radlmayr et al., 2014), there is a lack of systematic assessment of the effect of, for example, traffic complexity (due to varying degrees of road infrastructure elements or traffic density) in driver-paced conditions. Eriksson et al. (2015) found that drivers seek out different information depending on external factors, such as traffic complexity (in situations such as entering a city, driving on a motorway close to a junction and driving on an open motorway in low traffic) and time pressure (15, 30 and 120 seconds) in an online test. This lends support to the notion of letting drivers pace the transition process themselves, but it also raises questions regarding situations where a time budget must be used, such as (1) how does the required time budget change in different traffic situations, (2) what is the minimum time in such situations that allows drivers to still resume control in a safe manner (with minimum effects on driving performance) when traffic complexity increases and (3) does this change when the complexity element is caused by traffic density as in Radlmayr et al. (2014) or by different characteristics of the surrounding infrastructure? Moreover, do the information needs change when there are changes in infrastructure, for example, do drivers seek out information related to what the host vehicle is detecting as in Stanton et al. (2011), or does the driver want to know fuel status when on an empty motorway with no visible infrastructure whilst seeking out more tactical information when approaching an urban environment (Eriksson et al., 2015)?

8.4.2 THE ROLE OF DRIVER MONITORING

Whilst subjective measures of workload and situation awareness are a useful resource in the research phase, they are of little relevance in an automated driving system used by the general public. This means that future research should strive to design and implement a non-invasive driver monitoring system that allows objective assessment of driver state and behaviour. If driver state can be accurately determined through such non-invasive technologies, it would mean that certain elements of driver–vehicle interaction can be tailored to the situation.

For example, if the vehicle detects that the driver's attention is focussed elsewhere, it may issue an auditory alert, or even a vibration of the seat pad, to alert the driver of the need to attend to the system and direct attention to where relevant information is available, as exemplified in Chapter 7. Moreover, if information ascertaining driver state were to be fused with information regarding the external environment, that is, approaching a junction or a potential point where an alternate route may be embarked upon, the vehicle could then alert the driver to this alternative and actively accentuate that particular piece of information (i.e. display turn by turn navigation).

Moreover, if driver state can be accurately determined, the vehicle may tailor the take-over process to accommodate the increased time to resume control found in Chapter 4, which showed an increase in control transition time when drivers were engaged in a secondary reading task compared with passively monitoring the automation. If such a system were able to determine the type of task the driver is engaged in, and the time on task, the automated driving system could tailor the control transition process to accommodate the change in driver response-time. It is therefore recommended that further research into driver monitoring systems is conducted to ensure a safe, efficient transfer of control where the driver is reintegrated into the driving loop with the minimum amount of effort.

8.4.3 THE DISCREPANCY BETWEEN HUMAN FACTORS ENGINEERING AND VEHICLE DESIGN

Whilst conducting the research for Chapters 4 through 6, a discrepancy was identified between what was observed in the contemporary literature and the research findings presented in this book, with regard to control transition response-times compared with the sensors used in the automotive industry. It was found that whilst the contemporary literature shows that it takes drivers between 1.14 and 15 seconds to respond to a request to intervene, this book found that it takes drivers between 1.97 and 25.75 seconds. The drivers falling in the longer response-time part of this range give cause for concern as contemporary sensors lack the sufficient range to provide times longer than ~8 seconds due to physical limitations.

As an example, a driver would have 3.2 seconds to respond to a request to intervene if the stereoscopic camera from Autoliv (100-metre range; Autoliv, 2016) were to be used, and 5.11 seconds is the medium-range radar from Bosch (160-metre range; Bosch, 2016). If this were to be used, even with the longest-range sensor on the market (250-metre range; Bosch, 2009), the driver would have 7.98 seconds to respond to a request to intervene from an automated vehicle travelling at 70 mph (31.29 ms), assuming that the vehicle can identify the obstacle as soon as it is picked up by the sensor.

This discrepancy is worrying, as it has yet to be fully recognised in the academic discourse. It is therefore recommended that further research is carried out to either improve sensor ranges, reduce the time drivers need to resume control through adaptive support through human–machine interface solutions or increase the reliability of automated driving systems to such an extent that drivers are no longer required to act as a fallback when the system fails (SAE Level 3).

8.5 REFLECTIONS AND LIMITATIONS

8.5.1 EXPERIMENT DESIGN

Three of the chapters in this book utilise various parts of the same dataset collected from 26 participants in the Southampton University Driving Simulator; this has some methodological implications that must be acknowledged. When using one dataset for multiple analyses, the utmost care must be taken to acknowledge the challenges that come with such an approach. The primary concern is that the risk of a Type 1 error increases with the number of analyses run. A Type 1 error is a false detection of an effect that is not existent in the dataset being analysed. In order to avoid making such an error in this book, a lower alpha value was used to correct for multiple analyses. For the objective data, an alpha value of 0.01 was used for the main analysis, followed by the application of a Bonferroni correction, which is considered conservative, during post-hoc testing. For the subjective data, an alpha level of 0.05 was used for the main analysis, and post-hoc analyses were conducted using a Bonferroni correction.

Whilst this book could have avoided such corrections by utilising a larger number of experiments, there are some advantages to the approach used. Analysing the response-times to a request to transition between automated and manual control in Chapter 4 along with further analysis in Chapter 5 with data from a road testing vehicle did provide a validation not only of the algorithms presented in Chapter 3, but also the findings presented in Chapter 4 and the Southampton University Driving Simulator. In light of establishing task validity for the experimental set-up used in Chapter 4 through the analysis in Chapter 5, it was deemed appropriate to extend the analysis in Chapter 4 to include driving behaviour to assess the after-effects of automated driving (Chapter 6), as long as the appropriate precautions were taken.

Throughout this book, a substantial number of human participants took part in the experiments. Whilst this is common in Human Factors research in the automotive domain, it is also common to have to discard some participants, due to either missing data or simulator sickness. In the experiments presented in this book, only one participant was discarded due to simulator sickness, and due to robust data collection protocols, there was no missing data and no participants had to be discarded.

8.5.2 SIMULATORS

The generation of the results presented in this book primarily relied on the use of driving simulators; as such it is important to acknowledge some of the limitations of the use of driving simulation, as well as the advantages and what separates it from utilising real vehicles to conduct research.

As previously noted, 'The value of a simulator depends on its ability to elicit from the operator the same sort of response that he would make in the real situation' – Rolfe et al. (1970, p. 761). Consequently, it is of utmost importance to ensure that the simulator being used has been validated. Chapter 3 mentions two types of validity: physical and behavioural (Blaauw, 1982; Santos et al., 2005). Physical validity is achieved if the controls and driving environment correspond to that of a real vehicle (i.e. containing a steering wheel, pedals and other key components required for driving a vehicle) and an accurate representation of the vehicle dynamics in the simulated environment. Behavioural validity is somewhat more complicated as behavioural validity requires a driving environment that will elicit the same behaviour that the driver would exhibit in real driving conditions. This is not as straightforward to assess as physical validity as a driving simulator may not elicit a completely matched response (in terms of magnitude and direction of an effect), something referred to as absolute validity, but may instead generate an effect of a smaller or larger magnitude in the same direction, which is referred to as relative validity. Research has shown that, in some tasks, drivers tend to respond or act faster in real driving conditions compared with simulated driving (Eriksson and Stanton, 2017a; Kurokawa and Wierwille, 1990; Wang et al., 2010), which is an indication of relative validity. This difference could potentially be explained by the perception of greater risk in the on-road conditions (Carsten and Jamson, 2011; De Winter et al., 2012; Flach et al., 2008; Underwood et al., 2011).

Whilst this may seem like a limitation of driving simulators, there are several advantages to using the driving simulator that on-road testing cannot provide. Driving simulators allow researchers to assess driver reactions to new technologies to be measured in a safe, controllable and repeatable manner in a virtual environment without putting the drivers at risk of crashing (Carsten and Jamson, 2011; De Winter et al., 2012; Flach et al., 2008; Nilsson, 1993; Stanton et al., 2001; Underwood et al., 2011). Moreover, driving simulators enable researchers to test systems not yet available on production vehicles and inform design decisions on such technologies with minimal development cost.

8.6 CLOSING REMARKS

The overarching goal of this book was to establish how drivers resume control from automated vehicles in non-urgent situations, and how such transitions can be facilitated by allowing the drivers to pace the transition, or to change the predictability of a situation through different designs of the human–machine interface.

This book showed that it is possible to improve driver behaviour by changing different aspects (time and predictability) of what dictates 'level of control' in the Contextual Control Model (COCOM). This was demonstrated by letting drivers resume manual control from an automated vehicle without providing a time limit for the transition. This lack of a 'hard limit' for the transition of control showed substantial improvement in vehicle controllability when contrasted with reports in the contemporary literature utilising externally paced transitions with a short, fixed time budget. This research also highlighted the fact that there is more to the time it takes drivers to resume control from an automated vehicle than what can be represented by a single value (i.e. the mean or median response-time to a request to intervene). The book showed that there is a large variance both between and within individual drivers in how long they took to respond to a request to transition between automated and manual driving. This is a key finding as the research to date has focussed on the average driver, which in this book is argued to be problematic when designing the control transition process, as this would effectively exclude some drivers in the event of an externally paced transition designed for the average driver. Such a design could potentially give rise to fatal accidents and reduce public acceptance of highly automated vehicles.

It was also demonstrated that driver decision-making could be aided through Augmented Reality, pushing control from what in COCOM is referred to as 'Opportunistic' or 'Scrambled' control to a higher-level, 'Tactical' control. This finding supports the notion of designing the vehicle automation as a Chatty Co-Driver (as proposed in Chapter 2) providing the driver with continuous and task-relevant feedback.

It is hoped that the research presented in this book will encourage the designers of contemporary and future automated vehicles to consider the importance of the driver in automated vehicles (up to fully reliable SAE Level 4 automation in an operational design domain (ODD) that allows door-to-door automation) and to address the Human Factors identified in this book early on in the design phase.

REFERENCES

Autoliv (2016). Vision systems – Another set of 'eyes'. https://www.autoliv.com/ProductsAnd Innovations/ActiveSafetySystems/Pages/VisionSystems.aspx. Accessed on 14/09/16.

Blaauw, G. J. (1982). Driving experience and task demands in simulator and instrumented car: A validation study. *Human Factors: The Journal of the Human Factors and Ergonomics Society*, vol 24, no 4, pp 473–486.

Bosch (2009). Chassis systems control LRR3: 3rd generation long-range radar sensor. http://products.bosch-mobility-solutions.com/media/db_application/pdf_2/en/lrr3_daten-blatt_de_2009.pdf. Accessed on 14/09/2016.

Bosch (2016). Mid-range radar sensor driver assistance systems – Adaptive cruise control (ACC). http://products.bosch-mobility-solutions.com/en/de/_technik/component/CO_CV_DA_Adaptive-Cruise-Control_CO_CV_Driver-Assistance_2196.html?compId=2496. Accessed on 14/09/16.

Carsten, O., Jamson, A. H. (2011). Driving simulators as research tools in traffic psychology. In B. Porter (Ed.), *Handbook of Traffic Psychology* (vol 1). New York: Academic Press, pp 87–96.

De Winter, J. C. F., van Leeuwen, P., Happee, R. (2012). Advantages and disadvantages of driving simulators: A discussion. Paper presented at the Measuring Behavior Conference, Utrecht, the Netherlands, August 28–31, 2012.

Eriksson, A., Banks, V. A., Stanton, N. A. (2017). Transition to manual: Comparing simulator with on-road control transitions. *Accident Analysis and Prevention*, vol 102, pp 227–234.

Eriksson, A., Marcos, I. S., Kircher, K., Västfjäll, D., Stanton, N. A. (2015). The development of a method to assess the effects of traffic situation and time pressure on driver information preferences. In D. Harris (Ed.), *Engineering Psychology and Cognitive Ergonomics* (vol 9174). Cham, Switzerland: Springer, pp 3–12.

Eriksson, A., Stanton, N. A. (2017a). Driving performance after self-regulated control transitions in highly automated vehicles. *Human Factors*, vol 59, no 8, pp 1233–1248.

Eriksson, A., Stanton, N. A. (2017b). Takeover time in highly automated vehicles: Noncritical transitions to and from manual control. *Human Factors*, vol 59, no 4, pp 689–705.

Flach, J., Dekker, S., Stappers, P. J. (2008). Playing twenty questions with nature (the surprise version): Reflections on the dynamics of experience. *Theoretical Issues in Ergonomics Science*, vol 9, no 2, pp 125–154.

Gold, C., Korber, M., Lechner, D., Bengler, K. (2016). Taking over control from highly automated vehicles in complex traffic situations: The role of traffic density. *Human Factors*, vol 58, no 4, pp 642–652.

Grice, H. P. (1975). Logic and conversation. In P. Cole & J. L. Morgan (Eds.), *Speech Acts*. New York: Academic Press, pp 41–58.

Kircher, K., Eriksson, O., Forsman, Å., Vadeby, A., Ahlstrom, C. (2016). Design and analysis of semi-controlled studies. *Transportation Research Part F: Traffic Psychology and Behaviour*, vol 46, pp 404–412.

Kurokawa, K., Wierwille, W. W. (1990). Validation of a driving simulation facility for instrument panel task performance. In *Proceedings of the Human Factors and Ergonomics Society Annual Meeting*, October 1990, Santa Monica, CA: SAGE Publications.

Nilsson, L. (1993). Behavioural research in an advanced driving simulator – Experiences of the VTI system. In *Proceedings of the Human Factors and Ergonomics Society Annual Meeting*, vol 37, no 9, pp 612–616.

Parasuraman, R. (2000). Designing automation for human use: Empirical studies and quantitative models. *Ergonomics*, vol 43, no 7, pp 931–951.

Radlmayr, J., Gold, C., Lorenz, L., Farid, M., Bengler, K. (2014). How traffic situations and non-driving related tasks affect the take-over quality in highly automated driving. In *Proceedings of the Human Factors and Ergonomics Society Annual Meeting*, vol 58, no 1, pp 2063–2067.

Rolfe, J. M., Hammerton-Fraser, A. M., Poulter, R. F., Smith, E. M. (1970). Pilot response in flight and simulated flight. *Ergonomics*, vol 13, no 6, pp 761–768.

Santos, J., Merat, N., Mouta, S., Brookhuis, K., de Waard, D. (2005). The interaction between driving and in-vehicle information systems: Comparison of results from laboratory, simulator and real-world studies. *Transportation Research Part F-Traffic Psychology and Behaviour*, vol 8, no 2, pp 135–146.

Stanton, N. A., Dunoyer, A., Leatherland, A. (2011). Detection of new in-path targets by drivers using Stop & Go Adaptive Cruise Control. *Applied Ergonomics*, vol 42, no 4, pp 592–601.

Stanton, N. A., Young, M. S., Walker, G. H., Turner, H., Randle, S. (2001). Automating the driver's control tasks. *International Journal of Cognitive Ergonomics*, vol 5, no 3, pp 221–236.

Thellman, S. (2016). Social dimensions of robotic versus virtual embodiment, presence and influence. Student thesis, Linköping University, Linköping, Sweden.

Underwood, G., Crundall, D., Chapman, P. (2011). Driving simulator validation with hazard perception. *Transportation Research Part F – Traffic Psychology and Behaviour*, vol 14, no 6, pp 435–446.

Wang, Y., Mehler, B., Reimer, B., Lammers, V., D'Ambrosio, L. A., Coughlin, J. F. (2010). The validity of driving simulation for assessing differences between in-vehicle informational interfaces: A comparison with field testing. *Ergonomics*, vol 53, no 3, pp 404–420.

Author Index

Subject Index

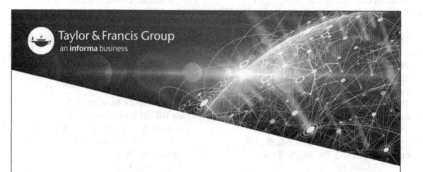